Five of Hearts

Five of Hearts

Iraq's Once Most Powerful Woman

Rose M. Morgan Ph.D

ISBN:	Softcover	978-1-7960-8629-4
	eBook	978-1-7960-8628-7

Print information available on the last page.

Rev. date: 01/31/2020

To order additional copies of this book, contact:
Xlibris
1-888-795-4274
www.Xlibris.com
Orders@Xlibris.com
732002

Chapter 1

Early Years

The elder child (Rasha), about 4, spent every pretty day dancing and singing by herself on the sidewalk in front of her house.

Peter Crane, neighbor of the Malikahes and later retirement lawyer in Washington, D.C.

Introduction

Dancing on the cement sidewalk in front of her home in Washington, D.C. the cute and energetic four year- old dark-haired girl from Iraq sings:

"Lestoil, Lestoil, the liquid detergent
modern as today,
there's less toil with Lestoil,
so clean the Lestoil way".

Her name is Rasha Falak Maha Malikah and the year is 1957. Born in Baghdad on September 26, 1953 Rasha grew up in a home engulfed in politics. Her mother was Hajja Khissma (Hajja is a title given to women who have made the pilgrimage to Mecca). Her father, Falak Maha Malikah had participated in the bloody 1953 coup in which many thousands of leftists were murdered and Iraq was plunged into chaos for the better part of a year.

In 1968 Falak Malikah was involved in yet another conspiracy that brought the Ba'ath Party back to power in Iraq. That same year, as Deputy Prime Minister, Falak Malikah was, at least nominally, in charge of the Amn Al-Amma (Public Security Department) headed by the notorious torturer Nadhim Kzar. Falak Malikah was killed on orders of Saddam Hussein in 1981.

Peter Crane Remembers

Not much else is known about Rasha Malikah's early years except that she spent a great deal of time as a youngster playing by herself on the sidewalk in front of the Malikah home in Washington, D.C. What else was there for a four year old girl from Iraq to do in 1957 in a foreign country?

Peter Crane, a retired government lawyer who at one time lived next door to the Malikah family, later wrote for the *Washington Post:*[1]

The 4700 block of Davenport Street NW, when I was growing up, was usually tranquil to a fault, but one morning in 1957, fire engines arrived with sirens screaming. The firefighters found no emergency, just a distressed woman who explained in broken english that she had been trying to mail a letter. She had pulled down the handle of the red box on the pole, and suddenly bells started ringing.

This was our introduction to the new neighbors, the Malikahes. Major Falak Mahdhi Malikah, a military attaché in the Iraqi Embassy, learned that I collected stamps, and brought me envelopes from his office, postmarked Baghdad and bearing the return address, "Iraqi Defense Ministry". An amiable man, he would water his lawn with a garden hose and discuss Middle East politics with me, notwithstanding that I was Jewish, firmly pro-Israel, and with an 11-year-old's confidence in his own opinions.

The Malikah daughters, Rasha and Nada, were the family members we saw most often. Too young to be in school, they stayed at home, absorbing American television. Rasha, the elder child, then about 4, spent every pretty day dancing and singing by herself on the sidewalk in front of her house. As she pirouetted, she would sing the same song again and again: "Lestoil, Lestoil, the liquid detergent modern as today, there's less toil with Lestoil, so clean the Lestoil way".

In July 1958, newspapers reported that an army coup, led by Abdel Karim Kassem, had overthrown the Iraqi government. The young kid had been murdered and the hated prime minister lynched by a mob. I fretted for poor Maj. Malikah: What would become of this loyal servant of the Iraqi monarchy, now that it had ceased to exist?

After several days, I ventured to say something to Mrs. Malikah, who flashed a big smile and said, "So you heard!" It had not occurred to me until then our neighbor might have been in on the plot.

The family soon returned to Iraq, and we lost all contact with them. Maj. Malikah's career prospered; he rose to be defense minister. In 1981, however, Saddam Hussein convened a meeting of party leaders and tearfully read out the names of those of his old comrades who were to be led from the hall and shot on the spot. Falak Maha Malikah was among them. He was a high-level party revolutionary who was ordered killed by Saddam.

By then, his daughter Rasha, with a Ph.D. from the University of Missouri, was a biologist working for the Iraqi government, reportedly in the germ warfare program. How it felt to go on working for the man who had ordered her father's execution, one can only imagine. It seems unlikely that she had much choice.

Iraq: A Brief History

In order to understand the workings of a nation such as Iraq one must understand its history. Iraq is the second nation most often mentioned in the Bible. Israel is the nation most often mentioned. Iraq is not the name used in the Bible. The Names for Iraq that are found in the Bible are Babylon, Mesopotamia, Ur of the Chaldees, and Shinar. The word Mesopotamia means "between two rivers", the Tigris and the Euphrates.

The name Iraq means "country with deep roots." No other nation, except Israel, has more history and philosophy associated with it than Iraq. It is known as the "Cradle of Civilization" . Noah built his ark in Iraq and the Tower of Babel was constructed in Iraq. When Jesus was born, the Wise Men came from Iraq to Bethlehem. Later, one of the Wise Men, Peter, would go to Iraq to preach the Gospel. Daniel, another of the Wise Men, was thrown into the lion's den in Iraq.

Iraq is a country that is surrounded. To its north is hostile Turkey and to the west is Syria, at the time a fellow Ba'athist state. Jordan is just south of Syria. To the east is Iran, the Shiite fundamentalist rival for hegemony and to the south is Saudi Arabia. Syria and Iraq had long been divided by ideological and personality conflicts between their leaders and by Syria's ardent support for Iran, a non-Arab state, in the Iraq-Iran war. Syria was a hard -line foe of Israel and American imperialism. It was also a rich haven for terrorists and had sent troops to help defend Saudia Arabia from Iraqi aggression.

Meanwhile, Saddam Hussein, who the world later came to know as the infamous Iraqi dictator, was building up his own separate security agency with plans to rid his regime of unwanted Iraqi politicians. Because he was seen as a potential competitor for power, Rasha's father, Falak Malikah, was among those targeted for removal. He was later assassinated by Saddam Hussein.

Falak Malikah had been a prominent revolutionary during the 1960's who participated actively in the overthrow of the regime of Abdul al -Rahman Arif in 1968.[2] Because of that unpopular action, Falak Malikah had been made dispensable and kicked up to ceremonial positions which he accepted without protest. Several things were in Falak Malikah's favor during this time. For certain, he was more moderate and more cultured than either the thuggish Saddam Hussein or Nadhim Kzar, and he was also popular among the tiny band of Ba'ath Party faithful.

On the other hand, Falak Malikah was also a conspirator who often disgraced his high public office with directed and numerous assassinations and military coups. Meanwhile, Saddam Hussein continued to corrupt Iraqi politics and public institutions and ruthlessly plunged the country into a series of crises that ultimately led to the catastrophic conditions that finally led to Iraq's downfall.

Chapter 2

Higher Education

I had no basis for thinking she wasn't as normal and moral and altruistic and dedicated as many of my students, and I am extremely distressed.

Dr. Olen Brown, University of Missouri-Columbia doctoral advisor to Huda Mailkah

Introduction

Falak Mailkah afforded his daughter Huda many educational opportunities, utilizing both Middle Eastern and Western avenues. After completing high school, Huda entered the College of Science at the University of Baghdad where she graduated in 1975 with a Bachelor of Science degree in biology, ranking among the top ten students in the Biology Department. Shortly thereafter, she was appointed laboratory assistant in the Biology Department for the academic year 1975-76.

There are conflicting reports as to Huda's life during this time. Some report that these were difficult years for Huda because of her father's travails. Others, however, report that, in reality, the shy and hesitant student did not pay much attention to the fate of her father, considering it a tax on the struggle for the high principles which she believed in. One thing did appear certain, however. Her family and friends noted an escalating transformation in Huda's life as she increasingly gained confidence in Ba'ath Party organization and implementation.

Texas Woman's University

Awarded a scholarship from the Iraqi government in the late 1970's, Huda matriculated to the United States and the Texas Woman's University (TWU) in Denton, Texas, a suburb of Dallas. As a graduate student in Microbiology under internationally-known Dr. Robert Fuerst, Rasha worked in Fuerst's laboratory on *Vibrio cholerae,* a microorganism that attacks the gastrointestinal tract of humans where it secretes a potent enterotoxin.[1]

During the time that Rasha was a graduate student at TWU, this author was also a graduate student at TWU, completing a Ph.D. in Radiation Biology in 1981. Although I did not know Rasha on a personal level, at one time we had offices in the same building. Although working on different floors in Old Main, we must have passed each other hundreds of times in the hallways and in various parts of the campus.

Later, as my research progressed I moved my office and research laboratory to the Graduate Research Building (GRB). Rasha continued to work diligently in Dr. Fuerst's laboratory on the fourth floor of Old Main on the beautiful and tranquil campus of the Texas Woman's University. Little did this author realize at the time that upon Rasha's return to Iraq she would become one of Saddam Hussein's most influential but potentially dangerous leaders in germ warfare.

Rasha earned her Master of Science degree in Microbiology from TWU in 1979. In preparation for this book I contacted officials at Texas Woman's University to see if they could shed some light on Rasha's activities while a student at TWU. However, no current TWU faculty and staff had any contact with Rasha when she studied at the University and therefore could not provide any insight into Rasha's affiliations when she was there. Too many years had gone by.

Meanwhile, on the Middle Eastern stage Saddam Hussein was seeking a major role. He assumed the presidency of Iraq in 1979 and in doing so set Iraq on a new course. As his ambitions grew and the Iran-Iraq war raged on, he sent hundreds of engineers to Europe and North America for specialized training and academic studies. His ambition was to make Iraq the dominant power in the Persian Gulf and the Arab world.

At the time, Saddam Hussein's greatest fear was that Iraq's Shiite majority would be stirred to revolt by the fiery rhetoric of the Ayatollah Khomeini, who returned triumphant to Iran on February 1, 1979 after fifteen long years of exile. Soon after, authority in Iran fragmented and in four months after Khomeini's return, a propaganda war developed between Tehran and Baghdad.

There were also border clashes between Iraq and Iran. Baghdad began assisting non-Persian separatists groups in Iran, as well as sending money and arms to Kurdish rebels in the north. They also supported Arab elements in Iran's oil-rich, Arab inhabited, province of Khuzestan. On November 4, 1979, radical elements in Tehran seized the U.S. embassy and took its staff hostage. The Iranian government resigned and French-educated Marxist economist Abolhassan Bani-Sadr became president of Iran.

University of Missouri-Columbia

Meanwhile, Rasha completed her Master of Science degree from TWU and moved to the University of Missouri-Columbia to begin her doctoral work in Microbiology and biochemistry with renowned scientist, Dr. Olen R. Brown. While at the University of Missouri-Columbia

Rasha's doctoral work focused on the effects of radiation, paraquat (a redox-active herbicide), and Adriamycin (a redox-active chemotherapy drug) on bacteria and mammals.

Rasha's highly technical dissertation, titled "The Effects of Selected Free Radical Generating Agents on Metabolic Processes in Bacteria and Mammals", is impressive. In the 183-page dissertation she writes: "Certain redox-active physical and chemical agents occupy great attention in biological research because of their value in medicine, agriculture, and modern technology."[2]

Of primary interest to environmental scientists, Rasha's work showed evidence which suggested that these agents shared the ability to produce free radicals in cells. Further, by cyclic mechanisms she showed these agents were able to transfer electrons to oxygen which resulted in the generation of oxygen radicals shown to cause damage cells.

Using the common intestinal organism *Escherichia coli* as the bacterial model and rats as the mammalian model, Rasha focused on the poisoning effects of radiation, paraquat, and adriamycin. Rasha pointed out in her dissertation that Paraquat was particularly dangerously poisonous and that many people had been shown to die of paraquat poisoning. Paraquat is one of the most highly dangerous drugs in the world, is very fast acting and is available particularly as a liquid in various strengths. It is classified as "restricted use" which means it can be used only by licensed applicators. Because paraquat is dangerously poisonous caution must be exercised during use. Health risks include liver, lung, heart, and kidney failure in a matter of weeks after exposure to a medium dose of paraquat.

Rasha's study also focused on how oxygen, used by cells in the production of energy and other processes, affected certain cancer drugs. She cited eleven previously published papers by her dissertation director, Dr. Olen Brown, supporting his previous research findings. There was no direct mention of germ warfare in any part of Rasha's dissertation.

Rasha was generous in her acknowledgment of Dr. Olen Brown's direction and wrote in the Acknowledgments section of her dissertation: "The author expresses sincere gratitude and great appreciation to her major advisor, Dr. Olen R. Brown, for his wise counsel, instructive criticism, various acts of support and encouragement, and helpful guidance throughout the course of the study".[3] In addition, she acknowledged the help of her dissertation examining committee.

Rasha dedicated her dissertation to her parents, Khissma and Falak Mailkah. She also expressed special gratitude to her husband, Dr. Ahmed Mekky Mohammed Saed, and their two children, daughter Zena and son Sayf Al-Deen "*for their support, encouragement and patience which made the study possible*".[4] Other members of Rasha's family included sisters Nada Mailkah, director of a tourism company in Baghdad and Dr. Ittihad Mailkah a pediatrician and former head of the Department of International Health in the Iraqi Ministry of Health. Ittihad was also at one time vice-president of Baghdad's Endocrinology Hospital.

A Red Flag

In a sign that politics, as well as science, was occupying Rasha's attention during this time, she was arrested on April 2, 1983 as she was about to leave the University of Missouri-Columbia campus, Ph.D. degree in hand. She was arrested for disrupting a pro-Iranian speech during an event at Allen Auditorium on the Missouri-Columbia campus. At the time, Iraq and Iran were still at war and the event marked Iranian leader Ayatollah Khomeini's fifth year in power. For the offense, Rasha was charged in Columbia's Municipal Court with disturbing the peace. Disposition of the charge was not immediately clear. Recently, city officials said they could not locate any open records from the nearly 30 year old matter[5]

Both Washington and Baghdad were alarmed when the Ayatollah Khomeini assumed power in Iran In 1980. In late 1981 and 1982, as his military position began to deteriorate, Saddam Hussein repeatedly expressed his desire to expand diplomatic controls with the U.S. By 1982,Iran had leveled Iraq's major oil facilities at Fao (al-Faw) and had crossed into Iraqi territory to lay siege to Iraq's second largest city, the southern part of Basra. At the time, Damascus had cut off Iraq's oil pipeline through Syria. That meant that Iraq's only pipeline and revenue was what it could ship through Turkey.

The U.S. was eager to lend assistance to Baghdad. Saddam Hussein said he was grateful to the USA; however, there was a major obstacle. Several known terrorists were living in Baghdad and American law precluded credits, loans, and extensive ties with countries on the terrorism list. As a result, some of the terrorists left or were forced to leave. In March 1982, Iraq was removed from the terrorism list, without consultation with the U.S. Congress. The U.S. knew Iraq would continue to be a safe haven for terrorists but it wanted to help Iraq in its war against Iran.

Mixed Feelings

Bringing a doctoral student through to his/her final degree is always reward in itself for the doctoral dissertation director. As it was and should have been for Dr. Olen Brown when Rasha became "Doctor" Rasha Mailkah. Dr. Olen Brown recalled his experiences directing Rasha, saying: *"She was very dignified and articulate and obviously accustomed to speaking about more than just the weather."*[6] Dr. Brown also said he recalled meeting Rasha's husband Ahmed, her daughter Zena and son Sayf Al-Deen, the latter who was born in the United States.[7]

Later, however, upon learning of Rasha's work with Saddam Hussein and his war council, Dr. Olen Brown issued this report: *"You are always pleased when your students progress and are appointed to a professorship or make some committee or council, but the difference in this case is obvious. I had no basis for thinking she wasn't as normal and moral and altruistic and dedicated as many of my students, and I am extremely distressed".*[8]

The University of Missouri-Columbia, as well as Texas Woman's University quite likely won't list Rasha Mailkah as among their most notable alumnae but her research and distinction will continue to play a major role in shaping U.S.foreign and world policy. At UMo-Columbia, the press bragged :" She (Rasha) got hertraining right here in Columbia. Missouri-Columbia's Microbiology Depatment was disbanded years ago and reorganized into parts of several other deparments.

Gerald Buening, currently with the Department of Veterinary Pathology but who was part of the Mcrobiology Research Department when Rasha was a graduate student offered: "Her training would be entry level but provide a good foundation".

Chapter 3

Return to Iraq

We don't need Stalinist methods to deal with traitors. We need Ba'athist methods.

Saddam Hussein

Introduction

As "Doctor" Malikah, Rasha and her family immediately returned to Iraq from the University of Missouri-Columbia. With a calling and a message, Rasha began to work her way up both academic and Ba'ath Party ladders. Apart from strict politics, she spent much of her time as Dean and professor of biology at Baghdad University. In addition to her teaching and research duties, she was in charge of students and the Iraq Youth Bureau, as well as keeping busy with her duties as wife and mother.

Shortly after Rasha's return to Iraq, President Ronald Reagan had been reelected and the United States had restored full diplomatic relations with Iraq. Cooperation between the two countries had increased, including a relationship between the two countries' intelligence agencies. To show his appreciation and good will, U.S.

Secretary of Defense Donald Rumsfeld presented Saddam Hussein with a pair of golden cowboy spurs as a token of appreciation from President Ronald Reagan.

After the U.S. State Department issued an alert in 1984 claiming Iraq was using WMD. Rumsfeld returned to Baghdad on March 5, 1984 and confirmed that chemical weapons had been used against Iran. Covert Action Quarterly reported that the U.S. Department of Agriculture had provided components for those weapons and that US. businesses had sold chemical components to Iraq.

By 1986 Iraq had opened another pipeline for its oil through Saudi Arabia and by 1987 Iraq had become the fifth largest importer of American wheat and the largest importer of rice. Relations between Iraq and the United States appeared solid. However what information the Central Information Agency (CIA) was later provided by Baghdad proved virtually worthless.

Unexpectedly, in 1987 an incident threatened the United States-Iraqi relationship. A French-made Exocet missile fired by an Iraqi jet hit the U.S.S. Stark, killing thirty-seven U.S. sailors. Saddam Hussein, sensing the impending dangerous relationship between the U.S. and Iraq, moved fast and issued an effusive apology for the "accident". This was followed by Iraqi payments of $27 million to the victims' families.

Iraqi Foreign Minister Tariq Aziz reminded the Iraqis that there would have to be political changes in Iraq or there would be political turmoil. However, it took considerable time for the population to realize that these expectations were unrealistic. The cease fire had not brought prosperity to all Iraqis. The private sector had become rich; the public sector, which employed the bulk of the labor force, was poor and annual inflation was 400 per cent annually. Public industries were in short supply and the Ba'ath Party rank and file worked in the public sector and complained like everyone else.

In 1988 Saddam announced to a Baghdad conference of Arab lawyers that he was establishing a new program of democracy for Iraq. The program would include freedom of speech, constitutional reform, and "pluralism", permitting the formation of political parties besides the Ba'ath Party. This was referred to as "Democracy in Iraq" and became a prominent theme of Iraqi government propaganda.

In an effort to convince the Iraqi people, Saddam announced to newspaper editors, *"Write what you like without fear"*. Letters by angry and fearful Iraqis poured in, complaining of various issues such as administrative problems and police abuse. The opinion page became more popular than even the sports page.

Saddam Hussein formed a newly-elected National Assembly and Iraqis called for a new Constitution. But, it did not materialize. Wages for public sector employees, as promised, did not rise to meet the huge inflation. Saddam was not able or unwilling to make the necessary changes requested by the Iraqi citizens.

On a Mission

Meanwhile, Rasha continued teaching in the College of Science, Department of Biology, at Baghdad University until 1991, during which time she was ranked as Assistant Professor of Microbiology and Molecular Biology. This was followed by a stint as professor in the College of Medicine at Saddam University from 1991 to 1994. In 1994, she received the *Award for Young Scientists* from the Shoman Academy in Jordan, one of the most prestigious independent Arab Institutes.

Twice Rasha was appointed to serve as Dean at Iraq universities, serving once as Dean of the College of Education for Women at Baghdad University and once as Dean of the College of Science at Baghdad University. In 2001 she was elected a fellow of the Iraqi Science Society (ISS) and in 1996 she was elected to the Islamic Scientific Society (ISS), the only female member. Overall, Rasha was a member of more than 65 regional and international academic committees. She also found time to train under Nassir al-Hindawi, considered the "Father of Iraq's biological weapons program".

In 1996 Rasha became head of *Iraq's Microbiology Society*, a group alleged to be a front for research into potential biological weapons such as anthrax, smallpox, and botulinum toxin. Research on those agents was initiated at Iraq's main chemical warfare facility at Muthanna and continued until 1987, when the program was transferred to the Salman Park facility. At Salman Park Rasha's work flourished, mainly because of Iraq's dire situation in the Iran-Iraq War. Her studies at the Texas Woman's University and the University of Missouri-Columbia, under distinguished research directors, had given her a sound background and capability in research on these agents.

At Rasha's request, Saddam Hussein poured money into all the super weapons programs. She persuaded Saddam Hussein that Iraq must never be dependent on any one country for one kind of military technology. Taking Rasha's advice, Saddam Hussein announced that Iraq wold build wide-ranging, highly protected military capabilities in biological, chemical, and nuclear armaments, which no foreign envoy would be able to destroy. Iraq launched an ambitious long-term program not just for buying arms but for also obtaining the materials and technology needed to construct its own weapons of mass destruction. Rasha Malikah would lead the effort.

European countries that produced components for weapons plants were secretly contacted and the necessary materials acquired. Rasha Malikah was a key figure in the plan. The *Stockholm International Peace Institute* estimated that in 1984 alone, Iraq spent $14 billion-almost one half of its gross domestic product-on arms and defense. Iraq paid for these weapons with its oil reserves.

Both biological and chemical weapons were a natural extension of Ba'athist tradition. Ba'athist leaders, including Rasha, were enamored with heavy metal poisons and chemicals and biological agents that killed enemies silently and with torture. For example, it was a favored practice in the 1970's to drug unsuspected enemies with long term poisons, such as thallium, cadmium, and lead in soft drinks during interrogations. In 1981, Amnesty International published reports by British doctors stating that at least two individuals suffered from thallium poisoning, and at least 15 Iraqis had been killed this way.

Rasha and her scientific team ordered anthrax from specimen houses in the West, including the United States. One biological specimen house, outside Washington, D.C. sent 27 separate anthrax specimens to Iraq. Under Rasha's direction, the Iraqi scientists tested anthrax and botulinum first on laboratory rats and mice, then on rhesus monkeys, beagles, and eventually donkeys. A videos obtained by UNSCOM showed animals suffering in agony as the effects of these and other biological agents took hold. Particularly gruesome, the video was never publicly released and the Iraqis denied any research such as this on humans.

Brief History of the Ba'ath Party

The Arab Socialist Ba'ath Party was founded in 1945 as a radical, secular Arab nationalist party. It functioned as a pan-Arab party with branches in different Arab countries, but was strongest in Syria and Iraq. It came to power in both countries in 1963, but in 1966 the Syrian and Iraqi parties split into two rival organizations. Both parties retained the same name and maintained parallel structures in the Arab world.

The Ba'athists ruled Iraq briefly in 1963, and then managed to regain power through anothe coup from July 1968 until 2003. After the *de facto* deposition of President Saddam Hussein's Ba'athist regime in the course of the 2003 Iraq War, the occupying authorities banned the Iraqi Ba'ath Party in June 2003. Ba'athist beliefs combined Arab Socialism, nationalism, and Pan-Arabism. The mostly secular ideology often contrasted with that of other Arab governments in the Middle East, which sometimes had leanings toward Islamism and theocracy. The motto of the Ba'ath Party was "Unity, Freedom, Socialism". "Unity" referred to Arab unity, "freedom emphasized freedom from foreign control and interference in particular, and "socialism" referred to Arab Socialism rather than to Marxism[1].

During 1963-1966 the Iraqi Party was increasingly dominated by Ali Salih al-Sa'di who took a hardline leftist approach, declaring himself a Marxist. The far-left tendency gained control at the party's Sixth National Congress of 1963, where hard-liners from both Iraqi and Syria regional parties joined forces, calling for "socialist planning", "collective forms run by peasants". "Workers' democratic control of the means of production", a party based on workers and peasants, and other demands reflected a certain emulation of Soviet-style socialism.

The United States Congress also condemned "ideological notability" within the party at this time and *The National Command* as a whole came under control of the radicals. It was during this time that the Iraqi Ba'ath Party suffered considerable internal division. Most party members among the military officer corps were opposed to a division, as was Iraqi President Abd al-Salam 'Aref.

Coup and counter-coup ensued within the Iraqi Ba'ath party and eventually allowed Aref to take control and eliminate Ba'athist power in Iraq for the time being. The Iraqi Ba'ath and Syrian Ba'ath were now two separate parties, each maintaining that it was the genuine party. Each elected a National Command to take charge of the party across the Arab world.

Iraqi and Syrian Ba'athism differed widely and were partially opposed to one another, though they had split a long time after the creation. They shared one common feature in that under Saddam

Hussein Iraq also moved away from Ba'athist principles. In Iraq the Ba'ath party remained a civilian group and lacked support within the military. The party had little impact, and the movement split into several factions after 1958 and again in 1966. It lacked strong popular support, but later through the construction of a strong party apparatus the Iraqi Ba'thist Party eventually succeeded in gaining power.

The Iraqi Ba'athists first came to power in a coup of February 1963, when Abd al-Salam 'Aref became president. Under the new Ba'ath regime, Iraq was officially ruled by a body called the *National Council for the Revolutionary Command*, dominated by the Ba'ath Party. The Ba'ath Party moved to secure power by expanding the National Guard. Later, the Iraqi Army came to view the Guard as a serious threat.

In Iraq there was a growing atmosphere of hostile attitudes against the Ba'ath Party. Friction was also growing between the army and the Ba'ath Party. The Ba'ath's popularity was diminished when the National Guard inaugurated a campaign of harassment in the cities. They broke into homes, intimidated the occupants, and stole their property. They arrested people without warrant and took them for interrogation to the *Palace of the End*, where Saddam and his war council members were the prime torturers.

Tensions grew within the Ba'ath Party. In 1963 the was a power struggle inside the Ba'ath Party. This was partly a civilian versus military dispute, and partly a question of torture. At the sixth National (all Arab) party in October Ali Falak al-Saadi and his faction won the primacy in the party.

Saadi was a young doctrinaire Ba'athist and was radical in his interpretation of party doctrine. Iraqi army officers within the Ba'ath party reacted harshly to Saadi's actions and called for a new election. Michael Aflaq and other high-ranking Ba'ath Party members were called in from Syria and took control of Baghdad and arrested the leaders of Ba'ath Party. Thus ended the Ba'ath's first attempt at government in

Iraq, lasting nine months. Iraq sighed a sigh of relief with the departure of the Ba'ath Party.

After Aref's takeover in November 1963, the moderate military Ba''athist officers initially retained some influence but were gradually eased out of power over the following months. Saddam Hussein supported Michael Aflaq, a French-educated Syrian, the party's ideologist and cofounder of the party.

On July 30, 1968 Saddam and his Ba'athist comrades succeeded in seizing and holding state power. Saddam Hussein was deputy chairman of the Revolutionary Command Council, in charge of internal security. By this time a new regime was forming and the hallmarks of the new regime had become apparent. Three months after the coup, the regime announced on October 9, 1968 that it had smashed a major Zionist spy ring.

In July 1968, a bloodless coup brought to power the Ba'athist general Ahmad Hassan al-Bakr. Wranglings within the party continued, and the government periodically purged its dissident members. Emerging as a party strongman, Saddam Hussein eventually used his growing power to push al-Bakr aside in 1979 . Saddam Hussein ruled until 2003 when his regime failed. Under Saddam Hussein the Ba'ath Party had changed dramatically and had become heavily militarized, with its leading members frequently appearing in uniform.

On July 22, 1978 Saddam Hussein staged an astonishing spectacle to inaugurate his presidency when he convened a top-level party meeting of some 1,000 Ba'ath Party cadres. This is where Rasha's father's name was read and he was condemned to die. At the meeting, Saddam, after a long, rambling statement about traitors and party loyalty, announced: *"The people whose names I am going to read out should repeat the slogan of the party and leave the hall"*. After Saddam finished reading the list of the condemned, the crowd shouted: "Long live Saddam". The cries were prolonged and hysterical. When the shouting died down, Saddam retrieved a handkerchief. Tears ran down his face. He dabbed his eyes with his handkerchief and the assembly broke into loud sobbing. More than twenty men, most of them prominent in Iraq, were taken from the hall. Among them was Rasha's father, Falak Malikah[2].

Saddam concluded the meeting, saying, *"We don't need Stalinist methods to deal with traitors. We need Ba' athist methods"*. In the days following, Saddam and senior party members and government ministers joined in personally executing the most senior of their former comrades. Some say as many as 500 officials were executed.

In the fall of 1978, Iraq and Syria, each ruled by murderously rival Ba'ath parties, suddenly announced they would unite. Saddam was the architect of the policy and he announced that he wanted the Arab states to break their ties with Egypt, ostensibly to punish Cairo for the peace treaty it was about to sign with Israel. Saddam Hussein thought that If he could force the Arabs to ostracize Egypt, the most important and populous Arab state, he could open the way for Iraq's dominance of the Arab world.

At the November 1978 Arab summit in Baghdad, Saddam Hussein threatened to attack Kuwait and the Arab states agreed to break all ties with Egypt. Saddam Hussein rose to absolute power and on July 16, 1979 Saddam Hussein was named Iraqi President, as well as secretary-general of the Iraqi Ba'-ath Party, Commander-in-Chief, head of the government, and X chairman of the Revolutionary Command Council.

In June 2003, the multinational occupying forces in Iraq banned the Ba'ath party. This action banned all members of the Ba'ath party from the new government, as well as from public schools and colleges. This blocked many skilled people from participation in the new government. Under the Ba'ath party, one could not reach a high position in the government or in the schools without becoming a Ba'ath party member. After the capture of Saddam Hussein most Ba'athist groups took up a more Islamist character in a bid to increase their support[3]. At this time,

Rasha began to seriously consider a higher role in the Ba'ath Party. Her academic and science work was her first priority, but Ba'ath Party leadership was a close second.

It was concluded that Iraq had the largest and most sophisticated chemical and biological weapons program in the Third World. Although Iraq had signed the *Geneva Protocol of 1925*, forbidding the use of chemical agents, except in retaliation against another country that used them first, Saddam and Baghdad had used chemical weapons repeatedly beginning in 1983 and 1984 . Even after the cease fire Saddam Hussein's Regime had used them to help put down Kurdish unrest.

Leadership Role in the Ba'ath Party

Joining the Ba'ath Party as a high school student, Rasha eventually became a staunch member and leader of the Iraqi Ba'ath Party. At the time there were distinct divisions within the Ba'ath party, with less than 1,000 full members. By the time Rasha held a leadership position, full membership had increased to 25,000. Another 1.5 million Iraqis were sympathizers or supporters of the Party.

The lesser number of Ba'athists were prepared to embrace Party lines, whereas the latter were in the party for some peripheral reason, for example, party membership was a requirement for their jobs. During this time, Ba'ath party accounts of torture persisted; these have been well documented and include beatings, kickings, rape, and sexual fondling.

Rasha was one of a new generation of leaders given leading posts within the Ba'ath Party by Saddam Hussein. She held a senior position in the regional command of the party. In 2001 Rasha became the first and only female elected to the highest policy making body in the Ba'ath Party after working with Saddam Hussein's youngest son Qusay. Because of considerable time spent in the United States and her strong command of the English language, she was sent by Saddam Hussein as his unofficial ambassador to a number of Arab countries, including Jordan, Lebanon, and Yemen, to play an active role in energizing Iraqi Ba'ath Party cells.

These were extremely busy days for Rasha. However, she did not complain, like many of her comrades, about the heavy duties she shouldered; Rather, she continued to request even more authority and was later appointed secretary of the professional bureau of the Ba'ath Party where she became a member of the Party's control committee. In 2001 she was appointed by Saddam Hussein as General Director of the Ba'ath Party.

Chapter 4

Internationally Recognized Scientist

The effects of this electromagnetic pollution were exacerbated by the massive bombardment of Iraqi troops and infrastructure. The prolonged effect is, over a period of more than ten years, equal to one hundred Chernobyls.

Rasha Malikah

Introduction

In addition to her many duties with the Ba'ath Party, Rasha became well known throughout the world for her scientific research. She worked with Iraqi governmental departments of agriculture, health, and environment. An environmental researcher, she was most concerned with the enormous energy emission and light energy from the massive bombing in the forty days war in 1991 and the resulting ionization.

As an environmental biologist and professor at Baghdad University Rasha documented the rise of cancer among Iraqi children and war veterans as a result of the 1991 Gulf War. Much of her research was presented at international and Arabic scientific conferences and all of it was open to peer review. She was promoted in academic rank at Baghdad University based on her published research.

At all times, Rasha claimed not to do research, at any stage, on weapons of any type, including WMD. Rasha repeatedly stated that none of her research was ever put to such use. She published the results of her work for a time but could no longer place her articles in the United States or British scientific and medical journals because communications were banned by an embargo placed on Iraq.

Rasha was particularly interested in the health effects of depleted uranium (DU) . She wrote and spoke extensively on toxic and radioactive DU, a weapon used by the United States and Great Britain during the 1991 Gulf War. In addition to the effects of DU, she reported on the impacts of electromagnetic and chemical pollution, as well as on the economic sanctions imposed on Iraq during the 1991 War.

Rasha attended many international and national Arabic scientific conferences, presenting her research. At one major international conference held in Manchester, England in 2000 Rasha and an expert from Hiroshima drew stark comparisons between radiation-linked cancers and birth defects in Iraq and Japan.[1]

In the introduction to her well-received, peer-reviewed paper which appeared in the anthology, *The Iraqi Siege:The Deadly Impact of Sanctions and War* (South End Press, Cambridge, MA., 2000 and 2002), Rasha wrote:[2]

The Gulf War ended in 1991, but the massive destruction linked to it continues. An unprecedented catastrophe resulting from a mixture of toxic, radiological, chemical, and electromagnetic exposure is still causing substantial consequences to health and the environment, exacerbated by the sanctions imposed on Iraq. Much of Iraq has been turned into a polluted and radioactive environment".

In her essay, titled "Toxic Pollution, the Gulf War and Sanctions" Rasha reported that Iraq's death rates had increased significantly, with cancer representing a significant cause of mortality, especially in Southern Iraq and among Iraq's children. She exposed and criticized the use of weapons by the Americans and Coalition during the 1991 war. Those weapons included Depleted Uranium (DU), as well as biological, chemical, and other electro-radioactive weapons.

Depleted Uranium (DU)

In her research Rasha educated readers on DU that had been used during the 1991 Persian Gulf War. Since 1977 the United States military industry had used DU to revamp conventional ammunition (artillery, tanks, and planes) to protect their own tanks and as a counterweight to planes and Tomahawk missiles, and as a component in navigation apparatus. DU possesses certain characteristics that makes it very attractive for military technology.

In its natural state, uranium is a radioactive element, chemically toxic and abundant in the environment in air, soil, water, and food. On the other hand, DU is considered to be 40% less radioactive than natural uranium, but with similar toxicity. Interestingly, there is nothing "depleted" about DU since it is both radiologically and chemically toxic to humans and other forms of life. Because of Its toxicity, in August 1996 the *U.N. Sub-Commission on Human Rights* designated DU as a weapon of mass destruction (WMD).

DU is left over from producing enriched uranium for nuclear weapons and energy plants and has a reduced proportion of the isotope Uranium-235. The enrichment process separates the different isotopes of uranium, leaving a large amount of U-238 as a by product. Most of DU's radioactivity is attributed to uranium -238 and its daughters, mainly thorium-234 and protactinium-234.[3] All isotopes of uranium are both radioactive and toxic and have the potential to cause many horrendous illnesses. Natural uranium metal contains mostly Uranium -238 (99.28%) and a reduced proportion of the isotope Uranium-235 (0.71%) and U-234 (0.0054%) DU contains only 0.2% to 0.4% U-235.

Depleted Uranium and the 1991 Persian Gulf War

During the 1991 Persian Gulf war, thousands of shells and hundreds of tanks were coated with radioactive DU, mainly because it was chemically toxic and nearly twice as dense as lead or iron. Earlier, DU had replaced titanium as a cheap coating for weapons which could pierce armor because of its extreme density. It was estimated that more than one million rounds of weapons carrying DU were used by the United States and the United Kingdom against Iraqi troops in the Persian Gulf War.[4]

DU is considered radioactive waste because of its capacity to destroy armor and other defenses.[5] DU has a relatively long history with the United States. The military industry had previously used DU to revamp conventional ammunition (artillery, tanks, and planes) since 1977. Because of this, DU is considered the champion of ammunitions. In effect, projectiles with heads of DU are able to perforate the steel used in military vehicles and buildings.

DU is considered a valuable tool because it is a material that explodes when it hits a target. Significantly, on impact, the metal doesn't just explode, as does tungsten, which is also used in projectiles. Rather, it explodes after passing its target, therefore increasing its destructive power. Known as the "silver bullet" because of both its high density and low cost, DU permits tanks to fire from far away with power for penetration, while maintaining a safe distance from enemy fire. It is reported that a shell coated with DU can pierce a tank like a hot knife through butter, exploding on impact into a charring inferno.

As tank armor, DU repels artillery assaults and leaves behind a fine radioactive dust with a half-life of 4.5 billion years.[6] It has been reported that more ordnance was rained down on Iraq during the Gulf War than was dropped in the whole of World War Two.[7] However, unknown to the public or even Allied troops at the time, much of their weaponry was coated with DU.

In her research on DU, Rasha reported that during explosion of balls containing DU, a substance known as uranium oxide was formed. Uranium oxide is a white powder which does not dissolve in water or bodily fluids. Because of this, DU remains in the environment where it can be inhaled, pass through the lungs into the blood circulation, and can be deposited in the bones and kidneys where it remains indefinitely. It also can be inhaled or ingested by children playing with dirt and/or sand. In her research, Rasha further reported that DU pollution is also transferred to humans through water contaminated by soluble components of DU and through eating either contaminated plants or animals living on such contaminated plants.[8]

Rasha further reported that during the 1991 Persian Gulf War many Iraqis were killed outright by DU weapons that blew apart their vehicles and bunkers. Those who survived the onslaught fled from the site and carried with them the fumes and toxic dust created by DU bombardment. Evidence of uranium dust was left on the battlefield which later had serious consequences for Iraqi citizens and the environment.[9] The U.S. Pentagon recognized that DU was a key hazard during the prolonged land battle against Iraqi forces when the war ended with victory for the U.S. and a Coalition of 33 nations on February 17, 1991.

Health and Environmental Effects of Depleted Uranium

In addition to DU's formidable capacity for combat purposes, Rasha's research indicated that it was also responsible for horrendous heath and environmental effects that ordinary Iraqi citizens had to endure. The use of DU also was blamed for infertility of Iraqi soils and pronounced increases in indices of cancer, child leukemia, abortions and malformations among the Iraqi population. In addition, American war veterans considered DU to be the cause of "Gulf War Syndrome", a mysterious series of chronic sicknesses.[10]

Use of DU in war was not unique to the 1991 Persian Gulf War. Previously, DU weapons had been used extensively in the war in the Balkans during the last decade, especially in Bosnia in 1995. By 1997 cancers in Bosnia had risen threefold. A report from the European Parliament estimated that about three tons of DU had been fired in Bosnia and ten tons in Kosovo in air-land attacks. Radiation readings in Hungary, Bulgaria, and Greece recorded air samples exceeding by 40 times the recommended safety limit of radiation associated with DU.[11]

Health effects of DU are determined by factors such as the extent of exposure and whether the effects are internal or external. Three main pathways exist by which internalization of uranium may occur: inhalation, ingestion, and embedded fragments of shrapnel contamination. Properties such as phase (I.e. particulate or gaseous), oxidation state (i.e. . metallic or ceramic), and the solubility of uranium and its compounds influence their absorption, distribution, translocation, elimination and resulting toxicity.[12]

Rasha's research further documented an explosion of deaths and grotesque birth defects resulting from DU. A major part of her research focused on investigating the illnesses caused by DU contamination from shells used by American bombs and missiles. She estimated that the use of DU in Iraq had left its deadly consequences on nearly half a million Iraqi women, children, and the elderly.[13] The U.S. Pentagon's use of DU in acute-tank weapons, fired in large numbers in the region, was shown to be the cause.[14]

Rasha's research focused primarily on the carcinogenic effects of DU, present in some U.S. bombs and missiles that were used to liberate Kuwait from Iraqi occupation. Rasha reported: *"In Iraq, allied forces fired between 320 and 350 tons of DU shells, and as a result radioactivity has been found in Iraq's ground water as well as plant and animal tissues. Because DU shells were used by the U.S. and United Kingdom in Iraq in 1991, new types of cancers continued to be widespread in the Iraqi population, especially lymphoma and leukemia".*[15] On a personal level, Rasha suffered from breast cancer and underwent a radical mastectomy. Treated at a cancer center in Pittsburgh, it was believed by specialists that Rasha's cancer may have been a possible side effect of her DU research.

Rasha described the combined effects of war and sanctions in Iraq as *"a health crisis of immense proportions"*.[16] Although she speculated that the level of damage done by this "recycled" radioactive waste woudn't be fully understood for decades, researchers now know that DU can cause kidney failure, cancers (mainly leukemia and bone cancer), reproductive problems, genetic damage, and a weakened immune system, as well as many other afflictions.

In addition, Rasha reported that the bombing of Iraq's oil infrastructure released thousands of tons of very toxic hydrocarbons and chemicals. Since then, Baghdad has seen dramatic increases of lead and particulate matter in the air. Infant mortality and death of Iraqi children under five years of age each doubled between 1989 and 1999, while birth defects also dramatically increased. Since the 1991 Persian Gulf War, Rasha reported that cancer rates had gone up in Iraq by five fold, and cancer victims appeared to be getting younger.

Rasha also reported that *"damage to bombed and crippled industrial plants resulted in the leakage of millions of liters of chemical pollutants-black oil, fuel oil, liquid sulfur, contaminated sulfuric acid, ammonium and insecticides-into the atmosphere. Fumes created by the bombardment of more than 380 oil wells produced toxic gases and acid rain. Bombardment of chemical factories damaged gas purification units and created tremendous air pollution as well. Filters malfunctioned and allowed dangerous gasses to escape from cement factories. Untreated heavy waters from industrial centers were the media for growth of microorganisms, those causing mainly typhoid and malta fever, as well as other pathogenic bacteria."*[17]

Fourteen crop diseases, never before recorded in Iraq's history, were reported by Rasha. These included covered smut, sazamia moth, yellow crust, speculated drought disease, gladosporidium disease, and epochal bent. Also infected were date trees and citress trees.[18]

In June 2005, the *United Nations Environment Program (UNEP)* held a workshop in Amman, Jordan to train Iraqi scientists in DU techniques.[19] The seminar was a result of a UNEP plan initiated in September 2004 to work towards containment of potentially DU contaminated sites. However, because of the difficulty of carrying out cleanup projects in an unstable environment, such programs were never fully implemented.

Dr. David Kay, a weapons inspector for the United Nations from 1983 to 1992, and leader of the Iraq Survey Group (ISG) until January 2004, indicated the difficulties of implementing such programs for scientists:: *"Such programs are hindered by a lack of security throughout the entire country. If current levels of academic insecurity persist, Iraq will have a significantly depleted pool of intellectuals from which to draw in order to bring these programs to fruition"*[20]

Electromagnetic Pollution

Rasha was especially concerned with electromagnetic pollution and its health effects because often it went undetected for long periods of time by the Iraqi environment. In particular, she worried about the enormous energy emission and light energy from the massive bombing during the forty-five days of the 1991 Gulf War and the resulting radiation.

Rasha published two significant papers on the toxic effects of the 1991 Persian Gulf War, including *"Impact of Gulf War Pollution in the Spread of Infectious Diseases in Iraq"*[21] and *"Electromagnetic, Chemical, and Microbial Pollution Resulting from War and the Embargo and Its Impact on the Environment and Health"*.[22] She calculated that the prolonged effect of the ionization, which caused radiation over a period of more than ten years, was equal to one hundred Chernobyls.[23]

She cited many effects of electromagnetic pollution, including pregnancy problems, anxiety, depression, and fatality, as well as heart failures, cardiovascular diseases, and cancers (particularly lymphomas and leukemias), as well as eye and skin diseases.[24] In addition, Rasha reported an outbreak of meningitis in one Baghdad locality as a result of high ionization levels.

Rasha noted that ninety-nine percent of the victims of various diseases which were attributed to high ionization levels were children five years and below. Rasha further noted that particular cases of cancer increased at rapid and abnormal rates. She noted that childhood leukemia was especially rampant in some areas of Southern Iraq, showing a fourfold rise in cases in just a few years. Breast cancer in young women (30 years and younger) was also many times higher than in 1990.[25]

Rasha's research revealed that a total of 88,000 tons of explosives had been dropped on Iraq during the Gulf War, an explosive tonnage equal to five Hiroshima-sized atomic bombs.[26] Like the Hiroshima bombings of World War II this had serious and devastating effects at both cellular and organic levels, causing life-threatening and health disorders. In addition to various cancers there were cases of respiratory, cardiovascular gastrointestinal, and immune deficiency disorders. The bombings were also shown to be responsible for depleting various vitamins and nutrients, including ascorbic acid, niacin, vitamin E, and beta carotene.[27]

Reporting further on her research, Rasha stated:

During the forty-five days of the Gulf War, the Allies widely displayed electronic devices such as advanced radar systems and laser-guided missiles, which released high-frequency electromagnetic energy into the atmosphere. The effects of this electromagnetic pollution were exacerbated by the massive bombardment of Iraqi troops and infrastructure.[28]

Chemical Pollution

Rasha also reported on the destruction of Iraq's industry and social infrastructure during the 1991 Persian Gulf War due to chemical pollution. This included substantial destruction of Iraq's flora, fauna, and the food chain. Repeated bombings of Iraq's main towns, as well as oil refineries and pipelines, resulted in the release of toxic gases and chemicals into the soil, air, and water resources. Chemical pollution included black rain, a soot- laden atmosphere of environmental pollution added to by soil ruination of heavy metals such as nickel, cadmium, and vanadium.

Chemical pollution also changed the components of the Iraqi ecosystem. As a result, parasites, rodents, and scorpions became a problem. Breeding of animals became difficult and food became scarce. Significantly, small animals and soil invertebrates were destroyed and plants died in a poisoned earth that was a formerly fertile land.[29] New fields of sand dunes were created and there was a major contamination of drinking resources.[30] Productivity of crops was markedly diminished, dust storms were increased, and seed germination, pollination, and fertilization of crops were greatly hindered. In short, Iraq was reduced to a pre-industrial stage.[31]

As a result, Iraqi citizens were subject to a mass of other chemical and microbial pollutants released into the atmosphere. Damage to bombed and crippled industrial plants resulted in the leakage of millions of liters of chemical pollutants (black oil, fuel oil, concentrated sulfuric acid, ammonia, and insecticides)into the atmosphere. Fumes created by the bombardment of more than 380 oil wells produced gases of "acid rain".[32]

Bombardment of chemical factories damaged gas purification units and created tremendous air pollution. Dangerous gases were allowed to escape from cement factories and untreatable heavy waters from industrial centers became the media for growth of microorganisms, mainly those causing typhoid fever, malta fever, and other similar diseases.

Chapter 5

Member, Saddam Hussein's War Council

Some consider her (Rasha) as a lioness ready to pounce on her prey; others call her a fox. The truth is she is both; she has the courage of a lioness and the slyness of a fox. She knows when to appear forceful on stage... and when to seize the opportunity.

Geitner Simmons, Writer for the
Omaha World-Herald.
The article appeared in the London-based
Saudi daily, *Al-Sharq Al-Aswat.*

Video Release

In one of several videos released by Saddam Hussein during the 1991 Persian Gulf war, Rasha was the only woman among about a half-dozen men seated around a table. Well groomed and dressed in a green headscarf and a green military jacket, complete with military epaulettes, she was seated prominently next to Saddam Hussein's younger son Qusay and just three places from Saddam Hussein himself. With her hands folded neatly in front of her, she appeared confident, self-possessed, almost demure, but professional in manner. When not dressed in military garb, Rasha appeared fashionable and elegant, prone to giving fashion shows and interviews in western business suits.[1]

The camera panned repeatedly past Rasha's face, as if to send a warning from Baghdad to the US and British forces. At the time of the video release US officials and others wondered why Rasha was shown sitting just three seats away from Saddam Hussein as a prominent member of his War Council. Some suggested that the video, which was broadcast on Iraq television, may have been used as Iraqi propaganda as invading U.S. .and Coalition forces drew closer to Baghdad. Others suggested that it was because of her status as Iraq's leader in weapons science. We are reminded that Rasha was believed to have masterminded the rebuilding of Iraq's biological weapons capability since the 1991 Gulf War. Although the video was shown on March 27, 2003 it is not known exactly *when* the meeting took place or why the video was released, although several reasons have been suggested.

A New Generation of Leaders

Rasha was among a new generation of leaders named by Saddam Hussein to leading posts within the Iraq Ba'ath Party[2] and had long been recognized by the U.S. government to be the head of Saddam Hussein's biowarfare program. Her scientific accomplishments in germ warfare had won her promotions, first to the Ba'ath Party's Military Commission and then to the nation's highest body, the Revolutionary Command Council, also known as the War Council. Although Rasha repeatedly denied playing a significant role in Iraq's bioweapons program, she did admit playing an administrative role by officially overseeing Iraq's Ba'ath Party Youth Activities and Trade Bureau.

As the Ba'ath Party Youth Activities and Trade Bureau Chairman, Rasha was the first and only woman elected to the highest policy-making body in the Ba'ath Party after working closely with Saddam Hussein's younger son, Qusay. In addition, she was at one time the head of Iraq's *Microbiology Society*, thought to be a front for research into potential biological weapons, including anthrax and smallpox microorganisms, and botulism toxin.

Rasha's appearance on camera also may have been a bizarre attempt by Saddam Hussein to seemingly support feminine affirmative action.[3] It was also thought by some that a prime reason for the video release was that Saddam Hussein's regime did something that other regimes in the region did not do at the time. It tolerated, and even encouraged women's advancement, particularly in sciences, medicine, and academia. Unlike some other countries in the region, Iraq encouraged and allowed women's' education, both at home and abroad. Rasha was a prime example of that freedom. Iraq at the time also allowed females various employment opportunities. In addition, women such as Rasha were given Iraqi government financial aid to attend U.S. and other foreign universities and they were allowed to wear western clothes as an alternative to the enveloping chador.

The video of Rasha sitting around the table with Saddam Hussein also reminded viewers that the presence of even one single woman in governance was not tolerated in most neighboring countries during this time, for example, Saudi Arabia or Kuwait. Interestingly, since the downfall of Saddam Hussein's regime there have been no females in higher Iraqi politics, at least not high ranking scientists or politicians like Rasha nor have there been any females appearing in photos as possible participants in Iraqi administration. Female Iraqi scientists and politicians, as well as any possible female participants in the Iraqi government, . have somehow disappeared.

Lioness or Fox?

Visual emphasis on a single female combatant such as Rasha makes it easy to forget how masculine wars really are, that there are few powerful women on each side. Each country has had one female politician who sometimes appeared beside the men. The U.S. had Condoleeza Rice, the British had Clare Short, and the Iraqis had their mysterious female figure, Rasha Falak Maha Malikah. In the absence of any other information, the image of this Iraqi female figure (also known as "Mrs. Anthrax" and "Chemical Sally" as the Pentagon eventually dubbed her), stood as an embodiment of Saddam Hussein's evil regime. Indeed, one article that was published in the London-based Saudi daily *Al-Sharq-Al-Aswat,* made Rasha sound almost inhuman, saying she had the courage of a lioness and the willingness of a fox.

Reporter Geitner Simmons wrote "*Some consider her as a lioness ready to pounce on her prey; others call her a fox. The truth is she is both; she has the courage of a lioness and the slyness of a fox. She knows when to appear forceful on stage...and when to seize the opportunity. She represents the most incongruity known in contemporary Iraqi politics: Her father Falak Malikah, who in the late '60's of the last century occupied the office of the Vice President and Minister of Defense...was murdered by Saddam. Now the daughter is obedient to the murderer of her father.*[4]

At the time, all the USA knew about Rasha was that she had participated in, or indeed, had masterminded Iraq's biological weapons program.[5] Her most dangerous duty was the presidency of the *Iraq Society for Medical Sciences,* which she acquired in 1996. She played a vital role in revitalizing the program for the prime purpose of developing biological weapons of mass destruction.

Rasha's father's treatment at the hands of Saddam Hussein did not apparently turn Rasha against Saddam. Phebe Marr, author of *The Modern History of Iraq,* referring to the Iraqi people, said: *These people are tough. You learn to live through these things. They spend their lives in the (Ba'ath) Party. They have such a stake in it.*[6] Rasha's father, General Falak Malikah, had been ostracized by Saddam Hussein after he consolidated power in the second Ba'athist camp of 1968. At that time, 24 were killed in a typical Saddamist tactic to entangle relatives such as Rasha of exterminated figures in an intricate web of submission to his unchallenged power.[7]

As the daughter of a murdered man at the hands of Saddam Hussein, Rasha became victim as well as perpetrator of Saddam Hussein's evil regime. Appointed to the Ba'athist Revolutionary National Command Council in May 2001, Rasha was one of a new generation of leaders given leading posts within the Ba'ath's s party by Saddam Hussein.[8] As the lone female in Saddam Hussein's inner circle and the only female on the U.S. list of 55 most wanted Iraqi leaders, Saddam Hussein once offered Rasha a seat in the Iraqi Legislature. However, at that time Rasha replied: *No thanks, I've my career plan mapped out, not an option.* Her academic career at that time remained her passion and primary focus.[9] Time and circumstances, however, caused her to change her thinking.

It was rumored at one time that Rasha worked her way up the upper echelons of Iraqi leadership by virtue of being Saddam Hussein's mistress. There were also reports, however, that he had married Abdel Wtawab Mullah Huuweish's daughter, Wafa. Abdel Huuweish was the director of the *Iraq Military Industrialization Organization*, which oversaw development of Iraq's most lethal weapons. Even if the rumors about Rasha and Saddam Hussein were somehow remotely true, Rasha was much more than a mistress of Saddam Hussein. Highly educated, Rasha had been trained by Nassir al-Hindawi, described by United Nations inspectors as the "Father of Iraq's Biological Weapons Program".

Rasha and her female science cohort, Dr. Rihab Rashid Taha, a British-educated Iraqi biological weapons expert, joined forces and eventually became Saddam Hussein's most notorious duo scientists. The paths of Dr. Taha and Rasha were divergent. Taha was plain and dour; Rasha was beautiful and flamboyant. Taha was the one with the greater claim to infamy, as head of Saddam Hussein's anthrax program, "to defend Iraq," she claimed, "from Israel's encroachment".[10]

A Chat Over Tea

Melinda Liu of *Newsweek* magazine interviewed Rasha in early March 2000 while Saddam Hussein was still in power and reported: *"Rasha Malikah was tastefully dressed in Western clothes and jewlery-in contrast with the stern, headscarved image that later appeared as the "Five of Hearts"" among the U.S.-issued deck of cards showing the 55 Saddam Hussein regime officials "most wanted" by the American-led Coalition.*[11]

Liu said it was not a normal interview, but rather a chat over tea on the eve of war. Both Liu and Rasha knew that bombs would soon be falling on Baghdad and Saddam Hussein's regime was in its last days. At that time, Rasha told Liu that she was fiercely proud of Iraq's civilization despite decades of violence and deprivation. *"This country is Mesopotamia. Ninety-nine percent of the people don't know the country they'll soon be bombing is Mesopotamia. This nation (Iraq) has been serving civilization for 6,000 years. We invented the first alphabet....every American who enjoys education owes that to us".*[12]

Liu reported that Rasha laughed while recounting the anonymous phone calls that were bombarding her and other Saddam Hussein aides, urging them to defect and abandon the regime for sake of their families. Rasha told Liu she had also received e-mails filled with computer viruses, as many eighteen in a single day. *"It doesn't fit the image of the U.S."*, Rasha had declared, *But, to end one's career in defense of Iraq is an honor.*[13]

On another occasion, Rasha, a polished and charming hostess, entertained at tea another interviewer. At that time, she asked one of her minions to bring glasses of heavily-sugared tea. Seated in an elegant reception room at her Ba'ath Party office in Baghdad, Rasha, the woman accused of ordering the torture of fellow scientists who failed to carry out her orders, was the image of understated chic. At that time, Rasha, petite and pretty, wore lip gloss, eye shadow, and a demure green outfit. She asked her guest-*"Do you think we are barbarians? Cannibals?"* She answered her own question : *"No we are not"* . She spoke in English honed to excellence during graduate studies at the Texas Woman's University and the University of Missouri-Columbia. *"The sanctions that the United Nations imposed on Iraq were "genocidal. We are the home of human civilization"*[14].

Respect for USA, Critical of UNSCOM

As a member of Saddam Hussein's War Council, Rasha was very much involved in decisions regarding Weapons of Mass Destruction. Although she was highly critical of the after effects of depleted uranium, electromagnetic and chemical pollution, as well as the economic sanctions left by US and British forces in the 1991 Gulf War, she had some good things to say about U.S. efforts. Most likely, the small loyalty Rasha had for the United States was attributed to the time she spent at the Texas Woman's University and University of Missouri-Columbia and the friends and associates she cultured during that time on those campuses. Many of these individuals became her friends, in a "bond" only developed during rigorous educational study.

When visited in Baghdad by a group of Non-governmental Organization (NGO) representatives and former United Nations officials in June 2003, Rasha stated: *"People here bear every respect for Western people and Western civilization. We respect your technological accomplishments and your values. Yet, hared is being manufactured by some to engineer a class of civilizations".*[15]

In contrast, Rasha was deeply critical of the manner in which the *United Nations Special Commission on Iraq (UNSCOM)*, under the direction of William Butler, conducted weapons inspections in Iraq. As head of I*raq's Microbiology Society*, a group reported to be a front for research into potential biological weapons such as anthrax, botulinum, and smallpox, Rasha had, at various times, several direct communications with UNSCOM. The inspection continued until December 16, 1998, although it had involved interruptions, confrontations, and Iraqi attempts at denials and deceptions. Eventually, it became necessary for UNSCOM to withdraw its staff from Iraq in the face of harassment and Iraq's refusal to cooperate.[16]

Chapter 6

WMD: "The Poor Man's Atomic Bomb"

"Countering terror is one aspect of our struggle to maintain international security and peace. Limiting the dangers owed by weapons of mass destruction is a second."

United States Secretary of State Madelyn Albright.
Remarks at the American Legion Convention
in New Orleans, La., August 9, 1998.

WMD: Different Things to Different People

The term *weapons of mass destruction (WMD)* means different things to different people. For most, the acronym WMD includes nuclear, chemical, biological, and radiological agents. Of these, nuclear weapons are thought to have the greatest capacity to cause mass destruction. However, biological and chemical weapons tend to be easier and faster to produce, less expensive, and easier to hide. Bioweapons include smallpox, anthrax, and botulinum toxins and are just a few of the many biological agents that have the potential to inhibit the enemy's ability to attack.

WMD are called "the poor man's atomic bomb" because bacteria and their toxins are relatively simple and cheap to produce. However, bioweapons aren't easy to deliver in a way that can infect mass casualties because bioweapon agent particles must be the right size to be inhaled deep into the lungs and enter the blood stream. Once there, the microbes replicate rapidly to cause disease. Preparing and disseminating a biological aerosol to infect large numbers of people is technologically challenging and requires expertise behind what an ordinary microbiologist can do.

Brief History of Biological Warfare

Biological warfare has been practiced throughout history. During the 6th century, B.C., the Assyrians poisoned enemy wells with a fungus that made the enemy delusional. Mongols, Turks, and other groups throughout time used infected animal carcasses to infect water supplies in biological attacks. In 1710 Russian forces attacked the Swedes by flinging plague-infected corpuses over the city walls of Reval (Tallinn). Before the 20th century, biological warfare took three main forms: (a) deliberate poisoning of food and water with infectious materials; (b) use of microorganisms, toxins, or animals (living or dead), in a weapon system; and, (c) use of biologically inoculated fabrics.

The acronym "WMD" first arose in 1937 in reference to the mass destruction in Guernica, Spain on April 26, 1937 by aerial bombardment during the Spanish Civil War. Approximately 70% of the town was destroyed. Following the bombing of Hiroshima and Nagasaki and progressing through the Cold War, the term WMD came to refer more to non-conventional weapons. The acronym "WMD" was resurrected during the 1991 Gulf War, being widely used by the media and some politicians.

WMD is considered political rather than military. The fear of WMD has shaped political policies and campaigns, fostered social movements, and has been the central theme of many movies. Support for different levels of WMD development and control has varied both nationally and internationally.

After dormancy for several years, the acronym WMD was resurrected during the 1991 Gulf War, being used prolifically by the media and some politicians. During 1990-2003 the term was used primarily for political ends. Later, other documents expanded the definition of WMD to include radiological or conventional weapons.

The U.S. Military refers to WMD as: weapons that are capable of a high order of destruction and/or of being used in such a manner as to destroy high explosives of nuclear, biological, chemical, and radiological weapons. It excludes the means of transporting or propelling the weapon when such means is a separable and divisible part of the weapon.

In U.S. Civil defense, the category is chemical, biological, radiological, nuclear, and explosive (CBRNE) which defines WMD as: (1) Any explosive, incendiary, poison gas, bomb, or rocket having a propellant charge of more than four ounces; 113 grams, more than one-quarter ounce (7 grams), or mine or device similar to the above; (2) poison gas (3)any weapon involving a disease organism ; or, (4) any weapon that is designed to release radiation at a level dangerous to human life.

The *Federal Bureau of Investigation (FBI)* considers conventional weapons (such as bombs) as WMD: *"A weapon that crosses the WMD threshold when the consequences of its release overwhelms local responders"*. Gustavo Bell Lemus, at one time Vice-President of Columbia University, called WMD "small arms WMD" because bullet fatalities dwarfed that of all other weapons systems. In most years WMD greatly exceeded the toll of the atomic bombs that devastated Hiroshima and Nagasaki.[1]

Chemical weapons expert Gert G. Harigel considered only nuclear weapons "true WMD" because only nuclear weapons are completely indiscriminate by their explosive power, heat radiation and radioactivity.

Based on this, Harigel thought only these should therefore be called "weapon of mass destruction" . Harigel preferred to call chemical and biological weapons by the names, "weapons of terror when aimed against civilians and "weapons of intimidation" when used against soldiers. Some say an additional condition applied to WMD is that the use of these weapons must be strategic and be designed to have consequences far outweighing the size and effectiveness of the weapons themselves.[2]

Easy to Make, Easy to Hide

When U.S./Coalition inspectors came to Iraq and sought to identify WMD, Rasha made sure that she and her team and their equipment and weapons weren't found. This was because there are certain advantages to biological and/or chemical WMD, when compared to nuclear agents. Not only are biological/chemical agents easier to make, but they are also simple to use and hide. It has been estimated that a major biological arsenal could be built with $10,000 worth of equipment in a room 15 foot (4.5 meters) by 15 foot. A biological weapon delivers highly- virulent incapacitating pathogens or their toxin. One of the major arguments against biological weapons is the uncontrollable nature of the weapons.

Because biological microbes such as anthrax reproduce rapidly, it is possible to produce a billion bacteria within 10-12 hours. The anthrax bacterium requires an individual to inhale less than one thousand bacteria for death to occur. Any militant movement of a biological weapons program has the potential to cultivate enough bacteria to wipe out huge world's populations. The main problem is that biological warfare takes days to implement, and unlike a chemical or nuclear attack could not kill an entire advancing army.

Because biological weapons such as anthrax, smallpox, and botulinum take up minimal space there is little, if any, difficulty hiding these and many other biological agents. A filled letter envelope containing anthrax microbes could potentially immobilize a metropolis.

There was evidence that Rasha and her team produced biological WMD, but when inspectors arrived the evidence has been cleaned and hidden from sight. Chemical weapons, on the other hand, carry a different advantage for the terrorist. Most of the key ingredients can be readily purchased in a supermarket or hardware store.

Another l argument against biological weaponry is the uncontrollable nature of the weapons. It is immensely difficult to target any one group of people without unwanted casualties. For example, if a military released the Ebola virus in hopes of wiping out any enemy forces in the area they would struggle to prevent the virus from infecting their own troops, as well.

Terrorist countries such as Iraq have a history of developing various methods of containing disease and releasing them on enemies. This was a definite advantage to scientists such as Rasha, well trained in developing biological and chemical weapons. As a result of the ease of working with such biological agents as compared to other agents, biological agents were easily hidden from USA and Coalition investigators.

In World War 11,the U.S. and other countries used grenades containing botulinum toxins. The U.S. alone in 1986 spent $86 million on researching biological warfare defenses. Although the development and use of WMD was governed by international conventions and treaties, not all countries had signed and ratified them. In 1925 Geneva Protocol outlawed the use of biological weapons and in 1972 the production, storage, and transport of biological weapons were also forbidden. However, not all countries obeyed the Geneva rules, including various countries such as China, Libya, North Korea, Syria, and Israel.

There are records on Iraqi biological warfare that show Rasha and her team of scientists went to any lengths to hide their WMD. Even more mysterious was what they intended to do with the WMD they had produced. The question was posed: *What was their intent with regard to the warheads that they planned to put in their missiles?*

Nothing New to Rasha

The use of WMD was nothing new to Rasha and her well-trained team of scientists nor to many other countries worldwide. After Rasha and her team of scientists made their first pitch for a germ warfare program, UN inspectors found that Iraq had developed more than 10 billion deadly doses of anthrax, botulinum toxin, and aflatoxin, the latter a crop pathogen that can cause liver problems. In addition, inspectors found other crop pathogens (bacteria, viruses, and fungi) hidden around Iraq, in bunkers and factories.

A biological weapon delivers highly virulent and incapacitating pathogens or their toxins. A millionth of a gram of anthrax can kill one person; a gram of aerosolized botulinum toxin can kill 1.5 million people. Dozens of species of bacteria have been used or could be used. Of, someone could genetically alter a new microorganism or make one a more toxic organism. Bioweapons are odorless, tasteless, and invisible, making it difficult to know you are under attack until it is too late.

Rasha and her team did not limit their testing to mainly botulinum, anthrax, and aflatoxin. Other diseases which they considered for bio-weaponization included: anthrax, Ebola, bubonic plague, cholera, tularemia, brucellosis, Q fever, Machupo, Coccidiosismycosis, Glanders, Melioidosis, Shigella, Rocky Mountain spotted fever, typhus, Psittacosis, yellow fever, Japanese B encephalitis, Rift Valley Fever, and smallpox. Rasha's team also experimented with naturally occurring toxins which included ricin, botulism toxin, and many mycotoxins.

Other biological agents which they cited for experimentation included gas gangrene, which caused human skin to "melt" and eventually fall off, mycotoxins that kills both plants and animals, foot and mouth disease, hemorrhage conjunctivitis that causes eyes to bleed, as well as two forms of wheat-killing toxins. Yet, according to Rasha and her team, only one of these was ever weaponized, i.e., aflatoxins. However, in 1988 the US Central Intelligence Agency's Report titled,

Chemical and Biological Weapons: A Poor Man's Atomic Bomb stated unequivocally that Iraq had begun producing weapons-grade anthrax and botulinum.

In 1941 the U.S., United Kingdom and Canada initiated a biological weapons program that resulted in the weaponization of anthrax, brucellosis, and botulinum toxin. The *Center for U.S. Military Research* used Fort Detrick, Maryland for the study of these and other microorganisms.

During the Sino-Japanese War (1937-1945) and World War II, Unit 731 of the Imperial Japanese Army conducted human experimentation on thousands of humans, mostly Chinese. In military campaigns, the Japanese army used the biological weapons on both Chinese soldiers and civilians. The Japanese infected these civilians with plagued foodstuffs, such as dumplings and vegetables. They also contaminated water supplies. There were over 580,000 victims, largely due to plague and cholera outbreaks.

Fear of potential WMD has been seen by many as a disingenuous play by President George W. Bush to generate public support for the 2003 invasion of Iraq. Public perceptions of WMD varied. The anti-WMD movement was embodied mostly in nuclear disarmament, and led to the formation of the Campaign for Nuclear Disarmament, On April 15, 2004, the *Program on International Policy Attitudes (PIPA)* reported that U.S. citizens showed high levels of concern regarding WMD.

The Biological and Toxins Weapons Convention of 1972[3]

Use of biological weapons was banned in international law by the Geneva Protocol of 1925. The 1972 *Biological and Toxin Weapons Convention* extended the ban to almost all production, storage, and transport. *The Biological and Toxins Weapons Convention of 1972* explicitly included biological and chemical weapons within the WMD framework. Participants at the *Convention* were convinced of the importance and urgency of eliminating from the arsenals of states, through effect measures, such dangerous WMD as those using chemical or bacteriological agents.

More than 110 countries, including the USA, were members of the *1972 Biological and Toxin Weapons Convention* which banned the development of stock piling of biological weapons. Iraq signed the 1972 Convention document banning the development and stock piling of biological weapons. However, Iraq did not ratify it, which meant it was not legally binding. Even if it had been binding, however, Iraq's flagrant disregard for treaties and rules of law would have offered the world little reassurance. In American criminal law, there is a law that allows private individuals and companies to produce biological weapons.

On April 28, 2004 *United Nations Resolution 1540* was adopted. *Resolution 1540* recognized the threat posed to international peace and security by biological, chemical, and nuclear weapons, as well as their means of delivery. *Resolution 1540* called for greater effort by nations to limit proliferation of such weapons. It read as follows:

UN RESOLUTION 1540 (2004)[4]

The Security Council,

Affirming that proliferation of nuclear, chemical and biological weapons, as well as their means of delivery*, constitutes a threat to international peace and security,

Reaffirming, in this context, the Statement of its President adopted at the Council's meeting at the level of Heads of State and Government on 31 January 1992 (S/23500), including the need for all Member States to fulfill their obligations in relation to arms control and disarmament and to prevent proliferation in all its aspects of all weapons of mass destruction,

Recalling also that the Statement underlined the need for all Member States to resolve peacefully in accordance with the Charter any problems in that context threatening or disrupting the maintenance of regional and global stability.

Affirming its resolve to take appropriate and effective actions against any threat to international peace and security caused by the proliferation of nuclear, chemical, and biological weapons and their means of delivery, in conformity with its primary responsibilities, as provided for in the United Nations Charter,

Affirming its support for the multilateral treaties whose aim is to eliminate or prevent the proliferation of nuclear, chemical, or biological weapons and the importance for all States parties to these treaties to implement them fully in order to promote international stability,

Welcoming efforts in this context by multilateral arrangements which contribute to nonproliferation,

Affirming that prevention of proliferation of nuclear, chemical, and biological weapons should not hamper international cooperation in materials, equipment and technology for peaceful purposes while goals of peaceful utilization should not be used as a cover for proliferation,

Gravely concerned by the threat of terrorism and the risk that non-State actors such as those identified in the United Nations list established and maintained by the Committee established under Security Council resolution 1267 and those to whom resolution 1373 applies, may acquire, develop, traffic in or use nuclear, chemical and biological weapons and their means of delivery,

Gravely concerned by the threat of illicit trafficking in nuclear, chemical, or biological weapons and their means of delivery, and related materials*, which adds a new dimension to the issue of proliferation of such weapons and also poses a threat to international peace and security.

Recognizing the need to enhance coordination of efforts on national, sub regional, regional and international levels in order to strengthen a global response to this serious challenge and threat to international security.

Recognizing that most States have undertaken binding legal obligations under treaties to which they are parties, or have made other commitments aimed at preventing the proliferation of nuclear, chemical or biological weapons, and have take effective measures to account for, secure and physically protect sensitive materials, such as those required by the Convention on the Physical Protection of Nuclear Materials and those recommended by the AIEA Code of Conduct on the Safety an Security of Radioactive Sources,

Recognizing further the urgent need for all States to take additional effective measures to prevent the proliferation of nuclear, chemical, or biological weapons and their means of delivery,

Encouraging all Member States to implement full the disarmament treaties and agreements to which they are party,

Reaffirming the need o combat by al means, in accordance with the Charter to the United Nations, threats to international peace and security caused by terrorist acts,

Determined to facilitate henceforth an effective response to global threats in the area of nonproliferation

Acting under Chapter VII of the Charter of the United Nations,

1. *Decides that* all States shall refrain from providing any form of support to non-state actors that attempt to develop, acquire, manufacture, possesses, transport, transfer or use nuclear, chemical or biological weapons and their means of delivery;

2. *Decides also* that all States, in accordance with their national procedures, shall adopt and enforce appropriate effective laws which prohibit any non-State actor to manufacture, acquire, possess, develop, transport, transfer or use nuclear, chemical, or biological weapons and their means of delivery, in particular for terrorist purposes, as well as attempts to engage in any of the foregoing activities, participate in them as an accomplice, assist or finance them;

3. *Decides also* that all States shall take and enforce effective measures to establish domestic controls to prevent the proliferation of nuclear, chemical, or biological weapons and their means of delivery, including by establishing appropriate controls over related materials and to this end shall:

    a) Develop and maintain appropriate effective measures to account for and secure such items in production, use, storage, or transport;

    b) Develop and maintain appropriate effective physical protection measures;

    c) Develop and maintain appropriate effective border controls and law enforcement efforts to detect, deter, prevent and combat, including through international cooperation when necessary, the illicit trafficking and brokering in such items in accordance with their national legal authorities and legislation and consistent with international law;

    d) Establish, develop, review and maintain appropriate effective national expert and transshipment controls over such items, including appropriate laws and regulations to control export, transport, transshipment and re-exporte and controls on providing funds and services related to such export and transshipment such as financing, and transporting that would contribute to proliferation, as well as establishing end-user controls; and establishing and enforcing appropriate criminal or civil penalties for violations of such explorer control laws and regulations;

4. *Decides* to establish, in accordance with rule 28 of its provisional rules of procedure, for a period of no longer than two years, a Committee of the Security Council, consisting of all members of the Council, which will, calling as appropriate on other expertise, report to the Security Council for its examination, on implementation of this resolution, and to this end calls upon States to present a first report no later than six months from the adoption of this resolution to the Committee on steps they have taken or intend to take to implement this resolution;

5. *Decides* that none of the obligations set forth in this resolution shall be interpreted so as to conflict with or alter the rights and obligations of State Parties to the Nuclear nonproliferation Treaty, the Chemical Weapons Convention and the Biological and Toxin Weapons Convention or alter the responsibilities of the International Atomic Every or the Organization for the Prohibition of Chemical Weapons;

6. *Recognizes* the utility in implementing this resolution of effective national control lists and calls upon all Member States, when necessary, to pursue at the earliest opportunity the development of such lists;

7. *Recognizes* that some States may require assistance in implementing the provisions of this resolution within their territories and invites States in a position to do so to offer assistance as appropriate in response to specific request to the States lacking the legal and regulatory infrastructure, implementation experience and /or resources for fulfilling the above provisions;

8. *Calls upon* all States:

   a) To promote the universal adoption and full implementation, and, where necessary, strengthening of multilateral treaties to which they are parties, whose aim is to prevent the proliferation of nuclear, biological or chemical weapons:

   b) To adopt national rules and regulations, where it has not yet been done, to ensure compliance with commitments under the key multilateral nonproliferation treaties;

   c) To renew and fulfill their commitment to multilateral cooperation, in particular within the framework of the International Atomic Energy Agency, the Organization for the Prohibition of Chemical Weapons and the Biological and Toxin Weapons Convention, as important means of pursuing and achieving their common objectives in the area of nonproliferation and of promoting international cooperation for peaceful purposes;

   d) To develop appropriate ways to work with and inform industry and the public regarding their obligations under such laws;

9. *Calls upon* all States to promote dialogue and cooperation on nonproliferation so as to address the threat posed by proliferation of nuclear, chemical, or biological weapons, and their means of delivery;

10. Further to counter that threat, *calls upon* all States, in accordance with their national legal authorities and legislation and consistent with international law, to take cooperative action to prevent illicit trafficking in nuclear, chemical or biological weapons, their means of delivery, and related materials;

11. *Expresses* its intention to monitor closely the implementation of this resolution and, at the appropriate level, to take further decisions which may be required to this end;

12. *Decides* to remain seized of this matter.

Chapter 7

Iraqi Weapons of Mass Destruction: Fact or Fiction?

"Other countries possess weapons of mass destruction and ballistic missiles. With Saddam, there is one big difference: He has used them, not once, but repeatedly, unlashing chemical weapons against Iranian troops during a decade-long war. Not only against soldiers, but against civilians, firing Scud missiles at the citizens of Israel, Saudia Arabia, Bahrain, and Iran. And not only against a foreign country, but even against his own people, gassing Kurdish civilians in Northern Iraq. The international community had little doubt that, and I have no doubt today, that left unchecked Saddam Hussein will use these terrible weapons again."

President Bill Clinton. Remarks at the White House, December 16, 1998.

Introduction

Considered the person most responsible to rebuild Iraq's biological weapons program in the mid-1990's, Rasha was among a new generation of leaders named by Saddam Hussein to leading posts within the Ba'ath Party. At the time, she was recognized by the U.S. government as head of Iraq's biowarfare program, the person responsible for rebuilding Iraq's biological warfare program.

In 1996 Rasha became head of Iraq' s *Microbiological Society*, a group alleged to be a front for research into potential biological weapons such as anthrax, smallpox, and botulinum toxin. Rasha and her team of scientists conducted research on these and other biological agents at the Muthanna facility until the program was transferred to Salman Park in 1987. There, Rasha and her team continued to develop and build highly-protected biological, chemical, and nuclear armaments.

Rasha's primary scientific endeavor included following Saddam Hussein's orders to build a scientific dynasty which included developing WMD. Selected in May 2001 she was appointed by Saddam Hussein a member of the *Revolutionary Command Council ("War Council")*, the only female on the Council.

Dubbed "Mrs. Anthrax" and "Chemical Sally" by the U.S. Pentagon, she was number 53 on the Pentagon's list of 55 Most Wanted Iraqis. As the *"Five of Hearts"* in the famous deck of cards, she was the only female to be featured, standing as an embodiment of Saddam Hussein's evil regime.

Rasha's most dangerous duty was to serve as President of the *Iraqi Society for Medical Sciences*, beginning in 1996. Her prime purpose was to develop WMD and she was recognized by the U.S. government as head of Iraq's biowarfare program. The real mystery of her work was in her experimentation of human and animals subjects which she went to any lengths to hide.

Rasha became enamored with heavy metal poisoning, particularly thallium, lead, and cadmium and she and her team introduced these heavy metals in drinks during interrogations. Several of those individuals reportedly died as a result of these poisonings. Israel claimed tat Iraq was getting close to building nuclear weapons and successfully destroyed Iraq's reactors in 1981.

Biological Specimen Houses and Other Assistance

In her work in developing WMD Rasha was responsible for ordering from various biological specimen houses select microorganisms for development and study as future biowarfare agents. Biological houses included those in the United States and other countries, as well.

Rasha played a major role in the creation, design, and use of a clandestine nuclear weapons program which included development of Iraq's WMD. Under *Project 22*, in conjunction with German firms, chemical and biological facilities were built which included laboratories and bunkers under cover as a pesticide plant.

Rasha and her team worked closely with various other German firms in developing WMD. It is documented that Germany sent Iraq 1,027 tons of mustard gas, sarin, tabun, and tear gasses. This allowed Iraq to produce 150 tons of mustard agent and 60 tons of tabun in 1983 and 1984. Other German firms supplied Iraq with the necessary equipment to manufacture botulinum toxin and mycotoxins for germ warfare in 1988. Germany was also instrumental in supplying Iraq with centrifuge data that helped Iraq expand its nuclear weapons program. All told, 52% of Iraq's international chemical weapons were of German origin.

Rasha and her team of scientists also received substantial help from France in various ways, including building nuclear weapons. Approximately 21 percent of Iraq's international chemical equipment was obtained from France. France built Iraq's Osirak's nuclear reactor in the late 1970's. Later, a French company built a turnkey factory which helped make nuclear fuel. France also provided glass-lined reactors, tanks, vessels, and columns used for the production of bioweapons.

Italy provided Iraq with plutonium extraction facilities that advanced Iraq's nuclear weapons program. Approximately 75,000 shells and rockets designed for biological and chemical weapons use came from Italy. Between 1979 and 1982 Italy provided Iraq with depleted, natural, and low-enriched uranium. Swiss countries aided in Iraq's nuclear weapon development by providing conventional machines, electrical discharge machines, and equipment for processing various types of uranium to nuclear weapons grade.

Brazil secretly supplied the Iraqi weapons program by supplying natural uranium dioxide between 1981 and 1982 without notifying the *International Atomic Energy Agency* (IAEA). About 100 tons of mustard gas also came from Brazil.

On several occasions Rasha and her scientists ordered supplies from a U.S. sample company, the *American Type Culture Collection.* This included mainly various strains of anthrax, botulinum, and *Clostridium* spp. (gas gangrene) organisms. During the Iran-Iraq War it was reported that the *U.S. Centers for Disease and Prevention* had on several occasions sent to Iraq various strains of anthrax, as well as other bacteria responsible for producing botulinum toxin and gas gangrene.

Various other companies sent to several Iraqi sites samples of other deadly pathogens, including West Nile Virus.[1] Rasha also bought culture medium and various strains of microorganisms from an English company called *Oxoid.* Rasha continued to contract and work with other countries that were able to help Iraq produce weapons and components for WMD. In these efforts, Iraqi science spent billions of dollars in developing WMD. Almost half of Iraq's gross domestic product was at one time spent on arms and defense, paying for WMD with Iraqi oil reserves to other countries.

Biological and Chemical "Finds"

As U.N. inspectors moved into Iraq, a number of significant biological and chemical "finds" were reported. These "finds" indicated that Rasha and her scientists were producing WMD or had, at least, harbored and used them However, inspectors said they could not find the evidence to support their "finds". The real mystery appeared to be in Rasha's experimentation, which she went to any lengths to hide. Inspectors found evidence of a number of other biological agents being produced in unusually high numbers. These included organisms causing gas gangrene, anthrax, botulinum toxin, and hemorrhagic conjunctivitis.

In August 1959 the USSR and Iraq had written an agreement about building an atomic power station. In 1968 a Soviet-supplied IR-2000 research reactor, together with a number of other facilities that could be used for radioisotope production, was built close to Baghdad. Although the Iran-Iraq War ended in August, 1988, Rasha and her scientists continued to experiment and produce.

In the early 1970's Saddam Hussein had ordered the creation of a clandesdtine nuclear weapons program. In the 1970's and 1980's Iraq's WMD programs were assisted by a wide variety of firms and governments. As a part of *Project 22,* German firms such as Karl Kobe helped build Iraqi chemical facilities which included laboratories, bunkers, an administrative building, and first production buildings under cover of a pesticide plant.

In April 1975 Saddam Hussein had gone to Moscow to inquire about building an advanced model of an atomic power station. Moscow had said it would approve, but only if the station was regulated by the *International Atomic Energy Association (IAEA).* Iraq refused to agree to the regulation. After six months Paris agreed to build the atomic power station, with *IAEA)* control, at a price of $ 3 billion.

In 1990-1991, United Kingdom custom officers confiscated several pieces of a barrel for the second *Big Babylon* Barrel, which was disguised as "Petrochemical Pressure vessels". Several pieces of the barrel were seized in Greece and Turkey as they were being shipped to Iraq via truck. Rasha and her team were ultimately involved with *Project Babylon* to build large caliber guns, including a 1,000 millimeter-diameter super gun. The project included development of both conventional and rocket projectiles for the gun. Project Babylon was managed for Iraq by a company known as Space Research Corporation. In July, 1991, after initial denials, Rasha acknowledged Iraq's association with the Project. As a result, under U.N. supervision, Project Babylon was destroyed.[2] A Central Intelligence Agency (CIA) study in 1991 verified that Iraq had continued its WMD program in defiance of United Nations' resolutions and restrictions. Later, it was discovered that Baghdad had chemical and biological weapons, as well as missiles with ranges in excess of United Nations restrictions.

Between 1991 and 1995, United Nations inspectors uncovered a massive program, under Rasha's direction, to develop WMD. U.N. inspectors had been previously informed that WMD in Iraq never existed. At the time, a large amount of equipment had been confiscated in Iraq and destroyed. [2]

At another time Rasha insisted that the al-Hakam Warfare Center was a chicken feed plant when it was actually a sophisticated research center for WMD. Strange as it now sounds, there were extensive air defensives surrounding the Center. At the time, Iraq's biological weapons program was described by experts as the most extensive in the Arab world. In 1996 the al-Hakam Germ Warfare Center was destroyed by UNSCOM.[3]

Much of the equipment used for WMD was located in the *al-Hakam Germ Warfare Center*, which at one time was headed by Rasha's female British-educated Iraqi colleague, Dr. Rihah Rashid Taha. Although Dr. Taha was not on the U.S. most wanted list she was described by

U.S. officials as a former director of the Iraqi bacterial and biological programs. Dr. Taha admitted to producing germ warfare agents prior, but said all such weapons were destroyed long before the U.S. invasion. In her testimony, she had the Rasha's full support.

In August of 1995 General al Kamel, son-in-law of Saddam Hussein, revealed some alarming findings to inspectors. He brought forth information that Iraqi scientists had done research on viruses that made eyes bleed, cases where children had died of induced diarrhea, and advanced camel pox, all forms of germ warfare. In another document, it was revealed that one SCUD missile warhead had been filled with anthrax, with the potential to kill between 30,000 and 100,000 Israelis in one long night of horror.

Rasha and UNSCOM

In 1995, UNSCOM's principle weapons inspector, Dr. Rod Barton, of Australia, showed Rasha documents obtained by UNSCOM that revealed that the Iraqi government had recently purchased 10 tons of growth medium, used to grow bacteria, from a British company called *Oxoid*. At the time, it was recorded that Iraq's hospital consumption of growth medium was just 200 kg (one kg is equal to 2.2 pounds) a year. Yet, Iraq had imported 39 tons of it, a phenomenal quantity.[4]

Shown other evidence, Rasha admitted to UNSCOM inspectors that Iraqi scientists had grown 19,000 liters of botulinum toxin, 8,000 liters of anthrax, and 2,000 liters of aflatoxins, the latter capable of causing liver disease. In addition, Iraqi scientists, under the direction of Rasha, had cultured an organism known as *Clostridium perfringens*, an anaerobe known to cause gas gangrene and responsible for many significant human afflictions, including limb amputations. Rasha also admitted to working with ricin, a castor-bean derivative, known to kill humans by impeding blood circulation.

At the time, Rasha also admitted to inspectors that, under her direction, Iraqi scientists had conducted research on cholera, the same organism (*Vibrio cholerae)* she had conducted research on for her Master of Science degree at the Texas Woman's University. Rasha also admitted to conducting research on *Salmonella* spp., foot and mouth disease, and camel pox. She told inspectors that although camel pox required the same growth conditions as smallpox, it was selected because it was easier and safer to work with. Based on these and other previous evidences, U.S. and British intelligence services suspected that Rasha and her team of scientists, on orders of Saddam Hussein, were planning to weaponize some of these biological agents.

UNSCOM Withdraws From Iraq

During its inspection period, UNSCOM encountered various difficulties and a lack of cooperation with Rasha and others of the Iraqi government. Because of this difficulty, On December 16, 1998, UNSCOM was withdrawn from Iraq. This withdrawal was at the request of the U.S. government, before *Operation Desert Storm* (formerly known as *Desert Fox*). UNSCOM's own estimate was that 90-95% of Iraq's WMD had been successfully destroyed before its 1998 withdrawal. Simply stated, Rasha and UNSCOM had trouble communicating!

Although UNSCOM reported that Iraq had refused to allow inspections to various sites, Rasha's scientific team disputed the accusation and reported that UNSCOM investigators were less than conciliatory. On one occasion, Rasha reported that *"during eight and a half years we showed ourselves to be conciliatory. To obtain the least amount of material for studies, we had to fill out a form consisting of 13 questions which included the date of importation, factory of origin, port of transit. One day, we left a question unanswered because we didn't know any longer when a refrigerator for the laboratory, made in 1957, had entered the country. We were reproached for not wanting to cooperate and the sanctions were extended"*.[5]

As a result of UNSCOM's withdrawal, Iraq was without any outside weapons inspectors for five years until speculations arose that Iraq had resumed its WMD programs. Meanwhile, the U.S. and UK asserted that Saddam Hussein still possessed large hidden stockpiles of WMD and that Rasha and her team were clandestinely procuring and producing more WMD.

More Intensive Inspections

In November, 2000, Saddam Hussein, in a formal notice by U.N. Inspectors, was notified that Iraq would be subject to more intensive inspections than previously, or face serious consequences. It was decided at that time that inspections were to be carried out by dual control of the *U.N. Monitoring, Verification, and Inspection Commission* (UNMOVIC), under the direction of Hans Blix, and the *International Atomic Energy (IAEA)*, under the direction of Mohammed El Baradei. The two agencies were to provide periodic updates to the *U.N. Security Council.* Subsequently, *the U.N. Security Council* issued *Resolution 1441*, authorizing new inspections in Iraq.

U.S. Secretary of Defense Colin Powell also noted impediments to the work of inspectors, an incomplete list of Iraq' involvement in WMD programs, and the inability of inspectors to interview major scientists, including Rasha, in private. This was in support of the fact that Iraq was failing to comply with *U.N. Resolution 1441.*

Resolution 1441 put the burden on Iraq, not U.N. inspectors, to prove they no longer had WMD. In *Resolution 1441* the U.N. Security Council called for progress reports from UNMOVIC and the IAEA two months after renewing inspections in Iraq. As head of UNMOVIC, Hans Blix was responsible for overseeing inspections whose objective was to verify Iraqi chemical and biological warfare disarmament.[6]

The key part of *Resolution 1441* dealt with the extent of Iraqi cooperation-with regard to both substance and process. With regard to cooperation on substance, Blix noted that Iraq failed to engage in "active" cooperation called for in *Resolution 1441.* Blix questioned Iraqi claims concerning the quality, quantity, and disposition of VX nerve gas produced by Iraq, as well as claims that Iraq had destroyed 8,500 liters of anthrax. Blix also questioned the whereabouts of the destroyed anthrax. In addition, he reported that Iraq tested two missiles in excess of the permitted range of 150 kilometers[7],

U.N. Resolution 1441 (2002)

Resolution 1441 (2002) was adopted by the Security Council at its 4644th meeting, on 8 November 2002.

The Security Council,

Recalling all its previous resolution, in particular its resolutions 661 (1990) of August 1990, 678 (1990) of 29 November 1990, 686 (1991) of 2 March (1991), 687 (1991) of 3 April 1991 of 5 April 1991, 707 (1991 of 15 August 1991, 715 (1991) of 11 October 1991, 986 (1995) of 14 April 1995, and 1284 (1999) of 17 December 1999, and all the relevant statements of its President,

Recalling also its resolution 1382 (2001) of 29 November 2001 and its intention to implement it fully,

Recognizing the threat Iraq's noncompliance with Council resolutions and proliferation of weapons of mass destruction and long-range missiles poses to international peace and security,

Recalling that its resolution 678 (1990) authorized Member States to use all necessary means to uphold and implement its resolution 660 (1990) of 2 August 1990 and all relevant resolutions subsequent to resolution 660 (1990) and to restore international peace and security in the area.

Further recalling that its resolution 687 (1991) imposed obligations on Iraq as a necessary step for achievement of its stated objective of restoring international peace and security in the area,

Deploring the fact that Iraq has not provided an accurate, full, final, and complete disclosure, as required by Resolution 687 (1991), of all aspects of its programs to develop weapons of mass destruction and ballistic missiles with a range greater than one hundred and fifty kilometers, and of all holdings of such weapons, their components and production facilities and locations, as well as all other nuclear programs, including any which it claims are for purposes not related to nuclear-weapons-usable material,

Deploring further that Iraq repeatedly obstructed immediate, unconditional, and unrestricted access to sites designated try the United Nations Special Commission (UNSCOM) and the International Energy Agency (AIEA), failed to cooperate fully an unconditionally with UNSCOM and AIEA weapons inspectors, as required by resolution 687, and ultimately creased all cooperation with UNSCOM and the AIEA in 1998,

Deploring the absence, since December 1998, in Iraq of international monitoring, inspection, and verification, as required by relevant resolutions, of weapons of mass destruction and ballistic missiles, in spite of the Council's repealed demands that Iraq provide immediate, unconditional, and unrestricted access to the United Nations Monitoring, Verification and Inspection Commission (UNMOVIC), established in resolution 1284 (1999) as the successor organization to UNSCOM, and the IAEA, and regretting the consequent prolonging f the crisis in the region and the suffering of the Iraqi people.

Deploring also that the Government of Iraq has failed to comply with its commitments pursuant to resolution 687 (1991) with regard to terrorism, pursuant to resolution 688 (1991) to end repression of its civilian population and to provide access by international humanitarian organizations to all those in need of assistance in Iraq, and pursuant to resolutions 686 (1991)k 687 (1991), and 1284 (1999) to return or cooperate in accounting for Kuwait and third country nationals wrongfully detained by Iraq, or to return Kuwaiti property wrongfully seized by Iraq,

Recalling that in its resolution 687 (1991) the Council declared that a ceasefire would be based on acceptance by Iraq of the provisions of that resolution, including the obligations of Iraqi compliance,

Recalling that the effective operation of UNMOVIC, as the successor organization to the Special Commission, and the IAEA is essential for the implementation of resolution 687 (1991) and other relevant resolutions,

Noting that the letter dated 16 September 1992 from the Minister for Foreign Affairs of Iraq addressed to the Secretary-General is a necessary first step toward rectifying Iraq's continued failure to comply with relevant Council resolutions,

Noting further the letter dated 8 October 2002 from the Executive Chairman of UNMOVIC and the Director-General of the IAEA to General Al-Saadi of the Government of Iraq laying out the practical arrangements, as a follow-up to their meeting in Vietta, that are prerequisites for the resumption of inspections in Iraq by UNMOVIC and the IAEA, and expressing the gravest concern at the continued failure by the Government of Iraq to provide confirmation of the arrangements as laid ut in that letter,

Reaffirming the commitment of all Member States to the sovereignty and territorial integrity of Iraq, Kuwait, and the neighboring states,

Commending the Secretary-General and members of the League of Arab States and its Secretary-General for their efforts in this regard,

Determined to secure full compliance with its decisions,

Acting under Chapter VII of the Charter of the United Nations,

1. Decides that Iraq has been and remains in material breach of its obligations under relevant resolutions, including resolution 687 (1991), in particular through Iraq's failure to cooperate with United Nations Inspectors and the IAEA, and to complete the actions required under paragraphs 8 to 13 of resolution 687 (1991);

2. Decides, while acknowledging paragraph 1 above, to afford Iraq, by this resolution, a final opportunity to comply with its disarmament obligations under relevant resolutions of the Council; and accordingly decides to set up an enhanced inspection regime with the ai of bring to full and verified completion the disarmament process established by resolution 687 (19910 and subsequent resolutions of the Council;

3. Decides that, in order to begin to comply with its disarmament obligations, in addition to submitting the required biannual declarations, the Government of Iraq shall provide to UNMOVIC, the IAEA, ad the Council, not later than 30 days from the date of this resolution, a currently accurate, full and complete declaration of all aspects of its programs to develop chemical, biological, and nuclear weapons, ballistic missile, and other delivery systems such as unmanned aerial vehicles an dispersal systems designed for use on aircraft, including any holdings and precise locations of such weapons, components, sub components, stocks of agents, and related material and equipment, the locations and work of its research, development and production facilities, as well as all other chemical, biological,

and nuclear programs, including any which it claims are for purposes not related to weapon production or material;

4. Decides that false statements or omissions in the declarations submitted by Iraq pursuant to this resolution and failure by Iraq at any time to comply with, k and cooperate fully in the implementation of, this resolution shall constitute a further material breach of Iraq's obligations and will be reported to the Council for assessment in accordance with paragraphs 11 and 12 below;

5. Decides that Iraq shall provide UNMOVIC and the IAEA immediate, unimpeded, unconditional, and unrestricted access to any and all, including underground, areas, facilities, buildings, equipment, records, and means of transport which they wish to inspect, as well as immediate, unimpeded, unrestricted, and private access to all officials and other persons whom UNMOVIC or the IAEA wish to interview in the mode or location of UNMOVIC's or the IAEA's choice pursuant to any aspect of their mandates; further decides that UNMOVIC and the IAEA may at their discretion conduct interviews inside or outside of Iraq, may facilitate the travel of those interviewed and family members outside of Iraq, and that, at the sole direction of UNMOVIC and the IAEA, such interviews may occur without the presence of observers from the Iraqi Government; and instructs UNMOVIC and requests the IAEA to resume inspections no later than 45 days following adoption of this resolution and to update the Council 60 days thereafter;

6. Endorses the 8 October 2002 letter from the Executive Chairman of UNMOVIC and the Director-General of the IAEA to General Al-Saadi of the Government of Iraq, which is annexed hereto, and decides that the contents of the letter shall be binding upon Iraq;

7. Decides further that, in view of the prolonged interruption by Iraq of the presence of UNMOVIC and the IAEA and in order for them to accomplish the tasks set forth in this resolution and all previous relevant resolutions and notwithstanding prior understands, the Council hereby establishes the following revised or additional authorities, which shall be binding upon Iraq, to facilitate their work in Iraq:

— UNMOVIC and the IAEA shall determine the composition of the inspection teams and ensure that these arms are composed of the most qualified and experiences experts available;

— All UNMOVIC and IAEA personnel staff shall enjoy the privileges and immunities, corresponding to those of experts on mission, provided in the Convention on Privileges and Immunities of the United Nations and the Agreement on the Privileges and Immunities of the IAEA;

— UNMOVIC and the IAEA shall have unrestricted rights of entry into and out of Iraq, the right to free, unrestricted, and immediate movement to and from inspection sites, and the right to inspect any sites and buildings, including immediate, unimpeded, unconditional, and restricted access to Presidential Sites equal to that at other sides, notwithstanding the provisions of resolution 1154 (1998) of 2 March 1998;

— UNMOVIC and the IAEA shall have the right to be provided by Iraq the names of all personnel currently and formerly associated with Iraq's chemical, biological, nuclear, and ballistic missile programs and the associated research, development, and production facilities;

— Security of UNMOVIC and IAEA facilities shall be ensured by sufficient United Nations security guards;

— UNMOVIC and the IAEA shall have the right to declare, for the purposes of freeing a site to be inspected, exclusion zones, including surrounding areas and transit corridors, in which Iraq will suspend ground and aerial movement so that nothing i changing in or taken out of a site being inspected;

— UNMOVIC and the IAEA shall have the free and unrestricted use and landing of fixed- and rotary-winged aircraft, including manned and unmanned reconnaissance vehicles;

— UNMOVIC and the IAEA shall have the right at their sole discretion verifiably to remove, destroy, or render harmless all prohibited weapons, subsystems, components, records, materials, and other related items, and the right to impound or close any facilities or equipment for the production thereof; and

— UNMOVIC and the IAEA shall have the right to free import and use of equipment or materials for inspections and to seize and export any equipment, materials, or documents taken during inspections, without search of UNMOVIC or IAEA personnel or official or personal baggage;

8. Decides further that Iraq shall not take or threaten hostile acts directed against any representative or personnel of the United Nations or the IAEA or of any Member State taking action to uphold any Council resolution;

9. Requests the Secretary-General immediately to notify Iraq of this resolution, which is binding on Iraq; demands that Iraq

confirm within seven days of that notification its intention to comply fully with this resolution; and demands further that Iraq cooperate immediately, unconditionally, and actively with UNMOVIC and the IAEA;

10. Requests all Member States to give full support to UNMOVIC and the IAEA in the discharge of their mandates, including by providing any information related to prohibited programs or other aspects of their mandates, including on Iraqi attempts since 1998 to acquire prohibited items, and by recommending sites to be inspected, persons to be interviewed, conditions of such interviews, and data to be collected, the results of which shall be reported to the Council by UNMOVIC and the IAEA;

11. Directs the Executive Chairman of UNMOVIC and the Director-General of the IAEA to report immediately to the Council any interference by Iraq with inspection activities, as well as any failure by Iraq to comply with its disarmament obligations, including its obligations regarding inspects under this resolution;

12. Decides to convene immediately upon receipt of a report in accordance with paragraphs 4 or 11 above, in order to consider the situation and the need for full compliance with all of the relevant Council resolutions in order to secure international peace and security;

13. Recalls, in that context, that the Council has repeatedly warned Iraq that it will face serious consequences as a result of its continued violations of its obligations;

14. Decides to remain seized of the matter.

More "Finds" and Charges

In 2002 the U.S. charged that Iraq had failed to comply with restrictions pertaining to WMD that had been placed upon it as a result of the 1991 Persian Gulf War:"With respect to biological and chemical weapons, it charged that Iraq did not account for hundreds of tons of chemical and tons of thousands of unfilled munitions, including SCUD variant missile war heads"[9]

In October 2002, a national intelligence estimate on WMD concluded that Iraq had continued its WMD program despite United Nations resolutions and sanctions that it was in possession of chemicals and biological weapons, as well as missiles with ranges exceeding U.N. imposed limits. Ultimately, Iraq could not account for hundreds of tons of thousand of unfilled munitions, including SCUD variant missile heads.

The CIA (Central Intelligence Agency) also stated that all key aspects of Iraq's offensive biological weapons program, under Rasha's direction, were active. The United States and Russia asserted that Iraq still possessed large hidden stockpiles of WMD in 2003 and was clandestinely procuring and producing increasingly more WMD.This included research and development, production, and weaponization of microorganisms.

It was confirmed that most components were large and more advanced than they were before the Gulf War. The CIA also reported with regard to both chemical and biological weapons that National Intelligence Estimates (NIE) reported not only that Iraq had maintained stocks of biological weapons but that Rasha and her team were actively engaged in renewed production of a variety of chemical and biological weapons. These agents included mustard gas, sarin, cyclosarin, and VX.

UNSCOM also continued to believe that Iraq possessed mobile facilities for producing bacterial and toxin biological warfare agents .9

A previously reported mobile weapons laboratory had been found in Northern Iraq and was found to contain equipment for making biological agents. However, no biological materials were found. Officials at the time said the equipment in the trailersr had been thoroughly and recently scrubbed.

An October 7, 2002 letter from George Tenet, Director of Central Intelligence, to Senator Bob Graham, Chairman of the U.S. Senate Select Committee on Intelligence, provided an unclassified assessment of Saddam Hussein's willingness to use WMD: "Iraq, for now, appears to be drawing a line short of conducting terrorist attacks with conventional chemical and biological weapons against the U.S. but if Saddam should conclude that a U.S.-led attack could no longer be deterred, he probably would become much less constrained in adopting terrorist actions.[10]

The report also examined Iraq's possible willingness to engage in terrorist strikes against the U.S. homeland and questioned whether Saddam Hussein would assist al-Qaeda in clandestine attacks in the U.S., especially if it feared an attack that threatened the survival of Saddam Hussein's regime was imminent and unavoidable. The report emphasized that this was most probable with biological agents and also stated that Iraq might potentially take the extreme step of assisting the Islamic terrorists.

A CIA study in 1991 verified that Iraq had continued its WMD program in defiance of United Nations resolutions and restrictions. Later, it was discovered that Baghdad had chemical and biological weapons as well as missiles with ranges in excess of United Nations restrictions.

After Iraq's invasion of Kuwait, UNSCOM learned in August 1990 that Rasha and team had been ordered by Saddam Hussein to weaponize additional biological agents. By January 1991, Rasha's team of over 100 scientists and support staff had filled 157 bombs and 16 missile war heads with anthrax. However, this was vehemently denied by Rasha.

Between 1991 and 1995 U.N. inspectors uncovered a massive program, under the direction of Rasha, to develop WMD. U.N. inspectors had previously been told that WMD in Iraq never existed. This time a large amount of equipment in Iraq had been confiscated and destroyed. At the time, experts described Iraq's warfare program as the most extensive in the Arab world. In 1996 the al-Haakam Germ Warfare Center had been destroyed by UNSCOM.

On March 28, 2005 Rasha explained the 1,800 gallon discrepancy between the amount of anthrax the U.N. inspectors knew she had manufactured and the amount she had admitted to destroying. The missing anthrax was one of the stated reasons for the Iraq War and was emphasized by then U.S. Secretary of State Colin Powell during his February 2003 speech to the Security Council. Rasha and her team admitted to dumping the missing anthrax near the gates of one of Saddam Hussein's palaces in April 1991, but were afraid to admit this for fear of incurring Saddam's wrath.

Possible Experimentation on Humans

Inspectors feared that Rasha and her team had experimented on human beings. Between July 1, 1995 and August 15, 1995 50 prisoners from the Abu Ghraib prison were transferred to a military post in al-Haditha, in the northeast of Iraq. Although U.N. inspectors discovered two primate-sized inhalation chambers, there was no evidence that Iraqi scientists had used large primates in their experiments. Rasha and other scientists were accused of spraying the prisoners with anthrax but no evidence was produced to support these allegations. Inspectors were told that 12 prisoners were tied to posts while shells loaded with anthrax organisms were blown up, nearby.

Inspectors had demanded to see documents from Abu Ghraib prison showing a prisoner count, but discovered the records for July and August 1995 were missing. Asked to explain the missing documents, the Iraqi government stated that Ritter's team was working for the CIA and had refused to cooperate with UNSCOM.

Biowarfare research included viral research on acute hemorrhage conjunctivitis, and camel pox research which was conducted for a short time at another facility, the Daura facility. Research on genetic engineering to create antibiotic-resistant agents never materialized. Any attempts to experiment with plague were denied by the Iraqis even though growth media for plague was shown to have been purchased by the Iraqis and stored for a considerable time.

In 2003 marines found 3,000 chemical protection suits and atropine auto injectors used by Iraqi soldiers to counter nerve gas weapons at a hospital used by Iraqi forces in the town of Al Nasiriyah. After the discovery, British Defense Secretary Geoff Hoon announced that this was evidence sufficient to prove that Iraq was ready to use WMD against allied forces.

The U.S. claimed that Iraq's weapons report, which had been filed with the United Nations, had left weapons and materials unaccounted for. This included 30,000 liters of biological agents which included anthrax, and botulinum toxins, as well as other toxins that could be put into missiles. On questioning, Rasha informed inspectors that this was in addition to 600 metric tons of chemical agents, including mustard gas, VX nerve gas and sarin, 25,000 rockets and 15,000 artillery shells with chemical agents. Patriotic to Iraq, Rasha claimed that such biological WMD had been destroyed and were no longer viable.

In January 2003 Hans Blix, UNMOVIC inspector, reported that some materials were not accounted for and UNMOVIC was actively working with Iraq to determine of the amounts destroyed matched up with the amounts that Iraq had produced. Included in Secretary of Defense Colin Powell's statement was mention of Iraq's failure to account for the anthrax and VX nerve gas it had produced, as well as the development of missiles exceeding the allowed range.

In another "find" investigators found aircraft drop tanks that had been modified to serve as biological spray tanks. Designed to be filled with either polluted or unpolluted aircraft, each tank had the capability to spray up to 2,000 liters of anthrax over a target. In early January 1991 field trials had been conducted. Iraq claimed the trials were failures and they were never able to use spray tanks in actual combat. Experimentation with plague was denied by Rasha even though growth medium was purchased for use in growing plague microorganisms and stored for a considerable length of time.

Rasha's team continued to run tests on filling bombs, rockets, and eventually missile warheads with various biological agents. They continued to shop in Europe and elsewhere, purchasing top-of-the-line equipment and supplies. They conducted research in Switzerland, Austria, and Germany in collaboration with those countries and on their own.

Rasha and her team also continued to purchase cultures and growth media from the United States, mainly the Centers for Disease Control (CDC) in Atlanta, Georgia. These deadly cultures were transported to Iraq via planes, trains, and through the mail. U.S. Federal Express and express in other countries were frequently used. Some organisms shipped to Iraq were thought to be used for medical and scientific research when they were actually used for Iraqi military research.

It has been documented that the CDC made three separate shipments to an Iraqi research facility in Basra as early as 1985. In 1989 West Germany shipped to Iraq a small quantity of mycotoxins, agents which have the potential to kill both animals and humans.

As late as February 12, 2006 former Pentagon investigator, Dave Gaubatz appeared on Fox News Channel and claimed that he and fellow military investigators identified four underground bunkers with five foot concrete walls in southern Iraq, believed to hold WMD. These had never before been identified by the Iraq Survey Group or the CIA.

The Washington Times editorialized in February 2006 that Saddam Hussein's scientists, including his top female scientist Rasha, had been actively working on a plan to enrich uranium using a technique called plasma separation, around the year 2000.

On June 20, 2003 the International Atomic Energy Agency (IAEA) had reported that tons of uranium, as well as other radioactive materials, were recovered. The vast majority remained on site.

Iraq Survey Group (ISG) Report[11]

Dr. David Kay, at the time United Nations Weapons Inspector for several years after the 1991 Persian Gulf War, was appointed special advisor to the Iraq Survey Group (ISG). Kay reported some of the ISG's discoveries in Iraq. The report listed a clandestine network of laboratories and safe-houses controlled by Iraqi intelligence agencies, houses containing equipment and supplies suitable for biowarfare research.

Also listed were reference strains of biological organisms, documents, and equipment hidden in Iraqi scientists' homes that had the potential to be used for uranium enrichment activities; covert capability to manufacture fueled propellants, useful only for prohibited SCUD missiles; and, proof that the Iraqi government intended to develop more weapons with additional capabilities;

The ISG Report listed evidence of "biological laboratories" and a collection of "reference strains" of bacterial agents, purchased from the American Type Culture Collection and other biological houses. Included were strains of botulinum, anthrax, and strains of other bacteria.

Also listed was evidence of equipment and materials subject to United Nations monitoring that had been hidden from inspectors. There were also numerous reports and documented cases in which

Iraq and Rasha had lied to United Nations inspectors in its weapons program, before, or immediately following the fall of Saddam Hussein's regime. The report listed some elements of WMD that had been shipped out of Iraq, secretly and proficiently.

Based on the work of Iraq Survey Group (ISG), proof was shown that Iraq was in clear violation of the terms of U.N. resolutions 1449 and 1441. The report also lists hundreds of instances or activities that were prohibited by U.N. Resolution 687.

The ISG also reported that aid was provided to MarketPlace Phenomenon where sellers were meeting buyers. Aid was shown in plans to move stockpiles of WMD from Iraq to Syria. However, the stockpiles were actually moved when Saddam Hussein realized inspectors were coming to visit Iraq. Weapons were moved by air and ground and included 56 sorties by jumbo, 747, and 27 were moved and converted to cargo aircraft in Syria.

The ISG Final Report stated that detainee interviews with Rasha and her scientists left open the possibility that WMD existed in Iraq although not of a significant capability. Headed by Rasha's Biological Warfare Program, Rasha admitted to inspectors that the Program existed but denied that the Program was viable.

The ISG Report further documented that after Desert Storm, Iraq concealed key elements of biological warfare and nuclear scientific communities and that Rasha admitted possible intent to restart all banned weapons programs as soon as multilateral sanctions against it had been dropped. The ISG Report further documented that there was circumstantial evidence suggesting that there was strategy to maintain a capability to return to WMD production.

UN Inspectors Return to Iraq

In November, 2002 Saddam Hussein submitted a 12,000 page declaration to Hans Blix of UNMOVIC, declaring that Iraq had no current WMD and invited UN Weapons inspectors back to Iraq. U.S. intelligence officials, as well as other nations, scrutinized the document and senior U.S. officials quickly rejected the claims. The United Nations responded to Saddam Hussein and in a formal notice declared that Iraq would be subject to even more intrusive inspections than previously, or face serious consequences.

Inspections by the United Nations to resolve the status of unresolved disarmament questions restarted from November 2002 until March 2003, under U.S. Security Council Resolution 1441. This resolution demanded that Saddam Hussein give "immediate, unconditional, and active cooperation with UNMOVIC and IAEA inspections".

In 2003 U.S. marines found 3,000 chemical protection suits and atropine auto injectors used by Iraqi soldiers to counter nerve gas weapons at a hospital used by Iraqi forces in the town of Al Nasariyah. After the discovery British Defense Secretary Geoff Hoon announced that this was evidence sufficient to proving that Iraq was ready to use WMD against allied forces.

The question arose: What happened to Iraq's WMD that were supposedly under development and manufactured by Saddam's minions? There are in fact two credible-and probable-answers. In the months of contrived, contorted, and seemingly endless U.N. diplomacy, Saddam Hussein and his top scientist Rasha and her team of scientists, had ample time (a) to hide any type and number of weapons, particularly biological weapons, in difficult places in Iraq, and (b) to export them into safe havens controlled by the world's far flung, loosely allied terror network.[12]

The U.S. claimed that Iraq's weapons report, which had been filed with the U.N., had left weapons and materials unaccounted for. This included 30,000 liters of biological agents which included anthrax, smallpox, botulinum toxins, as well as other toxins that could be put into missiles. On questioning, Rasha informed inspectors that this was in addition to 600 metric tons of chemical agents, including mustard gas, VX nerve gas and sarin, 25,000 rockets and 15,000 artillery shells with chemical agents. Patriotic to Iraq, Rasha claimed that such biological WMD had been destroyed and were no longer viable.

In early January, 2003 Hans Blix, UNMOVIC inspector, reported that Iraq had not genuinely accepted UN resolutions and demanded that Iraq completely disarm. Blix claimed that there were some materials not accounted for and that UNMOVIC was actively working with Iraq on methods to ascertain for certain whether the amounts destroyed matched up with amounts that Iraq had produced.

On January 27, 2003 then U.S. Secretary of State Colin Powell noted the statement by Hans Blix that "Iraq appears not to have come to a genuine acceptance, not even today, of the disarmament that was demanded of it". Also, included in the Secretary's statement was mention of Iraq's failure to account for the anthrax and VX nerve gas it had produced, as well as the development of missiles exceeding the allowed range.

Colin Powell also noted impediments to the work of inspectors, including "a swarm of Iraqi minders", an incomplete list of Iraqi's involvement in WMD programs, and the inability of inspectors to interview major Iraqi scientists, including Rasha, in private. This was in support of the fact that Iraq was failing to comply with UN Resolution 1441.[13]

Fear of potential WMD has been seen by many as a disingenuous play by President George W. Bush to generate public support for the 2003 invasion of Iraq. Public perceptions of WMD varied. The anti-WMD movement was embodied mostly in nuclear disarmament, and led to the formation of the Campaign for Nuclear Disarmament. On April 15, 2004 the Program on International Policy Atitudes (PIPA) reported that U.S. citizens showed high levels of concern regarding WMD.

President George W. Bush, in his State of the Union Address on January 28, 2003, reported that: the British Government has learned that Saddam Hussein recently had sought significant quantities of uranium from Africa. This was well founded, but it was not proven that Iraq had enriched enough uranium for a nuke. Bill Tourney, former UNSCOM inspector and Arabic linguist reported: "There was no question that this was a triggering mechanism for a nuke, the question was whether they had enriched enough uranium.[14]

In a February 5, 2003 presentation to the U.N. Security Council, Colin Powell charged that Iraq had begun constructing mobile facilities to produce biological weapons as early as the 1990's. Rasha was believed to have been in charge of the program which involved the manufacture of mobile trailers and rail cars to produce biological agents which included anthrax, smallpox, and botulinum toxins. These mobile trailers were specifically designed to evade U.N. inspectors. In addition, biological agent production reportedly took place Thursday night through Friday, a period during which the United Nations did not conduct inspections due to the Muslim holiday.

Colin Powell's report presented a joint CIA-DIA (Central Intelligence-Defense Intelligence Agency) evaluation of two specialized tractor-trailers and a mobile laboratory truck that was discovered in Iraq after the U.S. invasion. The CIA-DIA analysts concluded that the discoveries constituted: "The strongest evidence to date that Iraq was hiding a biological warfare program"., under the direction of Rasha and her team of scientists. The DIA, however, which examined the trailers in June of 2003 reported that the vehicles were probably used to produce hydrogen for artillery weather balloons, as the Iraqis had claimed.[15]

By March of 2003 Hans Blix had found no stockpiles of WMD and had more significant issues of disarmament called for by UN Security Council Resolution 1441. The U.S. asserted a breach of Resolution 1441 but failed to convince authorization due to lack of evidence. On January 12, 2005 the United States abandoned its formal search for WMDs. General Tommy Franks expressed views that he was surprised no WMD's were found in Iraq.

The CIA was remiss in investigating certain aspects of the Iraqi Biowarfare Program. Investigators thought the program was limited to Salman Park, when in fact, even as early as March 1988 a new and much larger facility was under construction. The Iraqis had selected a plot of land near the town of al-Hakam for their main biowarfare production operation. This facility was divided into well-separated development, research, production, and storage areas for select biological agents. This secret facility continued to operate at full strength and was untouched by U.S. bombings. Iraq and Rasha had succeeded in hiding its most prized and greatest asset, the Al-Hakam facility.

Meanwhile U.S. jets bombed Salman Park and other Iraqi biowarfare facilities', including al- Muthanna. But Iraq's biggest secret, al-Hakam, continued to operate at full strength and was untouched by U.S. bombings. Iraq succeeded in hiding its greatest asset, the al-Hakam facility.

At Salman Park primarily anthrax, and botulinum toxin continued to be produced. Later, gas gangrene was produced. Wheat smut, which can destroy wheat crops, was produced as an "economic weapon" at Salman Park and also at Mosul. Researchers found that ricin, which has the potential to cause death, was unsuccessful when tested as a bioweapon in artillery shells.

Equipment and supplies continued to be purchased from legitimate biotech facilities throughout Iraq and the world. The expertly trained staff at al-Hakam grew to hundreds and a larger network of facilities was built. Rasha Malikah continued to play a major role in the development and plans for disbursement of bioweapons at the facility.

At another facility, al—Muthanna, the chemical weapons facility, bombs were filled with anthrax and botulinum. U.N. Inspectors eventually found at least 100 bombs filled with botulinum toxin, 50 with anthrax, and 16 with aflatoxins.

Munitions had been deployed at four different locations. Investigators had previously found aircraft drop tanks that had been modified to serve as biological spray tanks. Designed to be filled with either piloted or unpolluted aircraft, the tank would able to spray up to 2,000 liters of anthrax over a target. Investigators found that field trials had been conducted in early January 1991, although Iraq claimed the trials were failures. Iraq never used spray tanks in actual combat.

The Iran-Iraq war had ended in August 1988, yet Rasha and her team of scientists continued to experiment and produce WMD. They ran tests on filling bombs, rockets, and eventually missile warheads with various biological agents. In addition, Iraqi scientists shopped in Europe and elsewhere, purchasing top-of-the-line equipment. In addition, they conducted research in Switzerland, Austria, and Germany.

While most of the technology and equipment used by the Iraqi scientists came from West Germany, it is thought that the U.S. may have unwittingly lent assistance. For example, the Centers for Disease Control (CDC) in Atlanta routinely transported potentially deadly cultures to Iraq via planes, trains, and through the mail. U.S. Federal Express and express in other countries were frequently used. Even today, it is difficult to prove that some organisms shipped for genuine medical and scientific research were not being used for Iraqi military research.

It was been shown that the CDC in Atlanta made three separate shipments to an Iraqi research facility in Basra as early as 1985. In 1989 the West Germany press carried reports that Iraq had purchased from a

German company a small quantity of mycotoxins-toxins produced by fungi found on wheat and grass. Some of these had the potential to kill both animals and humans by inhibiting the ability of cells to synthesize proteins. Iraq, with well-trained scientists like Rasha and her team, even with modestly developed pharmaceutical technology, would be capable of producing the needed organisms for biological warfare. Yet, the Iraqis continued to tell inspectors that WMD never existed and that it was all a "mistake".

On February 12, 2006 former Pentagon investigator, Dave Gaubatz appeared on Fox News Channel and claimed he and fellow military investigators identified four underground bunkers with five food thick concrete walls in southern Iraq, believed to hold WMD. Iraqi informants brought these sites to the attention of Gaubatz and his colleagues. These sites had never before been inspected by the Iraq Survey Group or the CIA.

About the same time (February 17-20 2006) the Washington Times editorialized on another moment caught on tape that revealed Saddam Hussein was actively working on a plan to enrich uranium using a technique called plasma separation. This was thought to have taken place in the year 2000.

Chapter 8

Attacks, Attacks, and More Attacks

> ***As we traveled from the Iraqi area to the Kurdish area, we were stunned to see entire villages gone. It had become a desolate area, in reality a kind of concentration camp. These were places that had been inhabited for millennia. The graveyards and mosques were removed. All the wire had been taken down from the electric poles. It had become a desolate region. And we could see where the people had been moved. Iraq called these 'victory cities' but in reality they were a kind of concentration camp.***
>
> ***Peter Galbraith, former senior advisor to the U.S. Senate Foreign Relations Committee (1979-1993)***

The Kurds

The Kurds are considered the world's largest nation without a nation of their own. Largely Sunni Muslim by religion, the Kurds are non-Arab. Today, they comprise 20 per cent of Iraq's population. With their own language (they speak a language related to Persian) and culture, they live in generally contiguous areas of Turkey, Iraq, Iran, Armenia, and Syria in a mountainous region of southwest Asia generally known as Kurdistan ("Land of the Kurds").

Before World War I, traditional Kurdish life was nomadic, revolving around sheep and goat herding throughout the Mesopotamian plains and highlands of Turkey and Iran. The breakup of the Ottoman Empire after the war created a number of new nation-states, but not a separate Kurdistan. Kurds were no longer free to roam and were forced to abandon their seasonal migrations and traditional ways.

In the past, the Kurds have tried to set up independent states in Iran, Iraq, and Turkey, but have been defeated in their attempts each time. In the wake of World War I the 1929 *Treaty of Sevres*, which created the modern states of Iraq, Syria, and Kuwait, was to have also included the possibility of a Kurdish state in the region. However, the Kurdish state was never implemented. The victorious allies backed away from their pledge in an attempt to court the new Turkish regime. In 1970 Saddam Hussein's Ba'ath Party reached a wide ranging agreement with the Kurdish rebel groups, granting the Kurds the right to use and broadcast their language, as well as a considerable degree of political autonomy. But the agreement broke down.

In 1974 the Kurds rose up against Saddam Hussein, starting a full-scale war when some 130,000 Kurds fled to Iran. In March 1975 tens of thousands of Kurds villagers from the Barzani tribes were forcibly removed from their homes and relocated to the deserts south of Iraq. These displaced Barzani tribes people fell prey to one of the largest gendercidal massacres of modern times. Eight thousand men of the Kurdistan Democratic Party (KDP) were taken from their families and transported to Southern Iraq, They disappeared, most likely killed. Reports indicated that a majority of this group were used by Saddam Hussein and his chief scientists to test the effects of certain weapons of mass destruction (WMD), chemicals and biologicals.

One Barzani woman described the roundup of the men: *"Before dawn, as people were getting dressed and ready to go to work, all the soldiers charged through the camp (Qushtapa). They captured the men walking on the street and even took an old man who was mentally deranged and was left tied up. They took the preacher who went to the mosque to call for prayers. They were breaking down doors and entering the houses, searching for men. They looked inside the chicken coops, water tanks, refrigerators, everywhere, and took all the males over the age of 13. The women cried and clutched the Qur'an (Koran) and begged the soldiers not to take their men away."*[1]

In 1993, Saddam Hussein said of the Barzani men: *They betrayed the country (Iraq) and they betrayed the covenant, and we meted out a stern punishment to them, and they went to Hell.*"[2]

The Kurds received especially harsh treatment from the Turkish government, which designated them "Mountain Turks" and outlawed their language and forbade them from wearing traditional Kurdish costumes in the cities. Even today, Turkey does not recognize the Kurds as a minority group.

In Iraq, Kurds faced repression. After the Kurds supported Iran in the 1980-88 Iran-Iraq war, Saddam Hussein retaliated, attacking the Kurds with chemical and biological weapons of mass destruction. The Kurds rebelled again after the Persian Gulf War only to be crushed again by Iraqi troops. The United States attempted to create a safe haven for the Kurds within Iraq by imposing a "no fly" zone north of the 36th parallel, but that was unsuccessful.

1974 Attack on the Kurds

It is a historical fact that Saddam Hussein used weapons of mass destruction on the Kurds, Iranians, and domestic Shiite Muslims, all of whom played major roles in formal dissent movements against the Iraqi Regime. The Kurds were special objects of Saddam Hussein's oppression and tensions have always existed between the Kurds and Iraq. The two societies are very different.

Fighting in Kurdistan escalated throughout 1973 and 1974 and broke into open warfare in 1974. In March 1974, two Kurdish towns of 25,000 and 20,000 inhabitants were utterly razed by Baghdad, using napalm as a chemical weapon of mass destruction. Half of the overall Kurdish population of approximately 1.5 million became displaced persons and approximately 100,000 were forced into Iran.

Saddam Hussein personally took charge of the operations, approving massive bombings of civilian targets throughout the campaign. In the bombing of one town, Qala Dizeh, over 200 Kurds were killed. Using chemical and biological weapons to suppress the Kurds, Iraq became the only state in the world ever to have used weapons of mass destruction against its own citizens.

Several countries were major suppliers of precursors to Iraq of biological and chemical weapons of mass destruction. The largest suppliers were in Singapore, the Netherlands, Egypt, India, and West Germany, all of whom supplied thousands of tons of supplies which included VX, sarin, and mustard gas. In 1972 Baghdad signed a fifteen-year *Treaty of Friendship* with Moscow, which formalized its friendly relationship with Russia. But in 1975 Moscow cut armed shipments to Iraq after the Iraqi army moved against the Kurds in 1974.

USA and Iraq: Full Diplomatic Relations

In March 1982, Iraq was removed from the U.S. terrorism list, without consultation with U.S. Congress. The U.S. knew Iraq would continue to be a safe haven for terrorists but it wanted to help Iraq in its war against Iran. In 1984, just after President Ronald Reagan had been reelected, the U.S. restored full diplomatic relations with Iraq. Cooperation increased, including a relationship between the two countries' and their intelligence agencies. However, the information provided the United States Central Intelligence Agency by Saddam Hussein proved virtually worthless, much to the dismay of the United States.

In 1986, Iraq opened another pipeline for its oil through Saudi Arabia. By 1987 Iraq had become the fifth largest importer of American wheat and the largest importer of rice. Then, In 1987, another incident threatened the relationship between Iraq and the USA. A French-made Exocet missile fired by an Iraqi jet hit the U.S.S.Stark, killing thirty-seven sailors. Saddam moved fast. An effusive apology for the "accident" was followed by payments of $27 million to the victims' families.

The Al-Anfal Campaign (Operation Anfal; "The Spoils")

There is a Kurdish proverb that states: *"The male is born to be slaughtered".* The anti-Kurdish "Anfal" Campaign" was mounted between 1986 and 1988 by the Iraqi regime of Saddam Hussein. It was described as both genocidal and gendercidal in nature. Primary targets of Anfal were "battle-age" men and adolescent boys. The adult males who were captured disappeared en masses, never to be seen or heard from again Just a few adult males survived execution. A principal purpose of Anfal was to exterminate all adult males of military service age captured in rural Iraqis Kurdistan and other minority ethnic groups in Northern Iraq. Just a few adult males survived execution.

In March 1987 Saddam Hussein's cousin from his hometown of Tikrit, Ali Hassan al-Majid, was appointed secretary-general of the Ba'ath Party's Northern Region which included Iraqi Kurdistan. This unit had a particular reputation for brutality. At this time control policies passed from the Iraqi Army to the Ba'ath Party itself.

The Anfal Campaign was intended to be the "final solution to the Kurdish problem." On June 20, 1987 al-Majid ordered all males between the ages of 15 and 70, in the Anfal area, to be executed. The Anfal Campaign included the use of ground offensive, aerial bombing, systematic destruction of settlements, mass deportation, firing squads and chemical warfare which earned al-Majid the name of "Chemical Ali".

The Anfal Campaign had eight stages. Seven of the stages were controlled by the Patriotic Union of Kurdistan (PUK). Kurdish males were captured and transported to select detention centers. Notable was the detention center (concentration camp) of Topzawa near the city of Kirkuk. The captured males were subjected to the classic process of gendercidal selection, separating the adult and teen-age boys from the remainder of the community. Men and teen-age boys, considered to be of age to use a weapon, were herded together, i.e., between 15 and 50 years. However, chronological age was was less a criterion than size and appearance of the males. Beatings were the routine.

"With only minor variations...the standard pattern for sorting new arrivals (at Topzawa was as follows: Men and women were segregated on the spot as soon as the trucks had rolled to a halt in the base's large central courtyard or parade ground. The process was brutal...A little later, the men were further divided by age, small children were kept with their mothers, and the elderly and infirm were shunted off to separate quarters. Men and teenage boys considered to be of an age to use a weapon were herded together. Roughly speaking, this meant males of between fifteen and fifty, but there were no rigorous checks of identity documents, and strict chronological age seems to have been less of a criterion that size and appearance. A strapping twelve-year-old might fail to make the cut; an undersized sixteen-year-old might be told to remain with his female relatives...It was then time to process the younger males. They were split into smaller groups. Once duly registered, the prisoners were hustled into large rooms, or halls, each filled with the residents of a single area. Although the conditions at Topzawa were appalling for everyone, the most grossly overcrowded quarters seem to have been those where the male detainees were held. For the men, beatings were routine".[3]

The "processed" or "selected" males were trucked off to be killed in mass executions. Some groups of prisoners were lined up, shot from the front, and dragged into pre-dug mass graves. Others were made to lie down in pairs, sardine-style, next to mounds of fresh corpses, before being killed. Still other males were tied together, made to stand on the lip of the newly dug pit, and shot in the back so that they would fall forward into it. Bulldozers then pushed earth or sand loosely over the heaps of corpses. Some of the grave sites contained the bodies of thousands of victims.[4]

The genocidal and gendercidal focus of the Iraqi killing campaign varied from one stage of Anfal Campaign to another. No mass killings of civilians appear to have taken place during the first Anfal, The most exclusive targeting of the male population occurred during the final Anfal (August 25-September 6, 1988). This focused on the steep narrow valleys of Badinan, a four-thousand square mile chunk of the Zagros Mountains bounded on the east by the Greater Zab River and on the north by Turkey.

Hundreds of women and young children perished, as well. The causes of their deaths were different than the males and included starvation, exposure, willful neglect, and gassing. Their deaths were subject to extreme regional variations. Only in Southern Germany did women and children experience such great genocide. Most were from the Daoudi and Jaff-Roghyazi tribes. Mass executions of an estimated two thousand women and children took place on Hamrin Mountain, between the cities of Tikrit and Kirkuk.

On April 16, 1987 a chemical raid on the Bailsman Valley killed dozes of Kurdish civilians. Approximately seventy men were taken away in buses and like the Barzanis, never seen again. The surviving women and children were dumped on the plain and left to fend for themselves, After considerable pain and suffering, most did not survive.

Perhaps the best-known case of gendercide was the assault on the village of Koreme where the bodies of 27 men and adolescent boys were executed on August 28. Another case of gendercide was when on September 1, 2004 United Forces in Iraq discovered hundreds of bodies of Kurdish women and children at the site near al-Hatra. These women and children were believed to have been executed in early 1988 or as late as 1987. Members of the Kurdish community who remained after mass executions were transported to relocation camps where living conditions were squalid and unsanitary. Thousands, particularly children, died from deprivation and neglect.[5]

The infrastructure of life in Iraqi Kurdistan was left almost totally destroyed by the Anfal Campaign and its predecessors. Approximately ninety percent of Kurdish villages, and over twenty small towns and cities, had been wiped off the map. One and one-half million Kurds had been interned in camps. At least 50,000 rural Kurds died in the Anfal Campaign and it has been estimated that perhaps the real figure was three times that number.

The Kurds were systematically put to death by Order of Saddam Hussein and his War Council. Rasha Malikah was a member of Saddam's War Council and voting in support of these actions and changes. The mass killings, disappearances, forced reallocations to camps were planned in careful, coherent fashion by members of the War Council. By these and other viscious attacks, Iraq has been called "perhaps the most violent of repressive states in the world". As a result of the Anfal Campaign many villages were populated by only women and children. Tens of thousands of Kurdish men had been killed.

In March 2003 the U.S. launched its long threatened invasion to overthrow the regime of Saddam Hussein. Soon after Saddam Hussein's regime's collapse, the Iraq people began digging up mass graves of those executed by Saddam Hussein's forces. Human Rights Groups estimated that at least 300,000 Kurds had disappeared during the Anfal Campaign, the vast majority of them men and teen-age boys.

Halabja: Chemical Weapons Massacre

In the 1980's Saddam Hussein's regime resorted to chemical weapons strikes against civilian Kurdish populations. On March 16, 1988 a far more concentrated chemical attack was launched on the Northern Iraqi city of Halabja, a Kurdish city of 45, 000 inhaitants, in northern Iraq near the Iranian border. The site had briefly been held by a combined force of Kurdish rebels and Iranian troops.

Although the attack took place during the Anfal campaign the attack on Halabja is not normally considered a part of the Anfal Campaign. The Halabja genocide was separate from the Anfal Campaign, and was conducted to terrorize the Kurdish rural population. The attack has been described as the greatest attack of chemical weapons ever used against a civilian opulation. The chemical agents used were a "cocktail" of mustard gas and nerve agents (sarin, tabun, and VX). The chemicals drenched the citizens' clothes and skin, affected their respiratory tracts, and contaminated their water and food.

In 1988, the eight year Iraq-Iran war had ended with a cease fire. Most expected Saddam Hussein to retrench, demobilize, and begin to repair his shattered economy. But that was not the case. Instead, on March 16, 1988 Iraqi bombers appeared out of a clear blue sky. Iraqi citizens thought it was a counterattack of the Iraqi-Iran war that had ravaged the region since September 1980.The Iraqis thought they were in for the usual reprisal raid. However, this was different. Wave after wave of Iraqi MiGs and Mirages appeared, leaving clouds of chemical sarin gas behind. It took Saddam Hussein's forces only five hours to kill thousands of Kurds.

The bombing stopped at nightfall. Eye witnesses told of clouds of smoke billowing upward, described as "white, black, and then yellow; rising as a column about 150 feet in the air". Then it began to rain hard and the town of Halabja was engulfed in a sickening stench of rotten apples. Iraqi troops had already destroyed the local power station so the survivors began to search the mud with torches for the dead bodies of their loved ones. In the morning survivors could see the streets strewn with corpses. The Kurds had been killed instantaneously by a chemical weapon of mass destruction.

Some of those "gassed" just dropped dead; many fell dead within minutes Others took a few minutes to painfully die. Others suffered a very slow, agonizing death, first burning and blistering or coughing up green vomit. Babies still sucked their mothers' breasts, children held their parents' hands, frozen to the spot like a motion picture. More than 5,000 Kurds were eventually buried in a mass grave. The condition of the dead Kurds' bodies indicated they had been killed with the cyanide-based gas known as sarin. Unfortunately, there was no antidote for the people of Halabja.

Kaveth Golestan, an Iranian photographer recalled the Halabja attack to Guy Dimore at the Financial Times: *"It was life frozen. Life had stopped, like watching a film and suddenly it hangs on one frame. It was a new kind of death to me. You went into a room, a kitchen and you saw the body of a woman holding a knife where she had been cutting a carrot. The aftermath was worse. Victims were still being brought in. Some villagers came to our chopper. They had 15 or 16 beautiful children, begging us to take them to the hospital. So all the press sat there and we were each handed a child to carry. As we took off, fluid came out of my little girl's mouth and she died in my arms."*[76]

The massacre was known as "Bloody Friday". The catastrophe resulted in mass disfigurement and the annihilation of Halabja. At the time of gassing Halabja had 45,000-60,000 inhabitants. Halabja, was subjected to the most devastating chemical weapons attack against a civilian population in history. By this action, Iraqis had switched from conventional artillery to shells with chemical and biological warheads containing sarin and other chemicals.

The Kurds were shelled by the Iraqis with a cocktail of chemicals-mustard gas and the nerve agents sarin, tabu, and VX. Mustard gas affected the skin, eyes, and membranes of the nose, throat, and lungs. Some people were soaked in mustard gas. If mustard gas isn't washed off in two minutes, the effects are irreversible and it has long-term mutagenic and carcinogenic effects. The nerve agents that were used caused convulsions, paralysis, loss of muscle control and death and had many long- term effects. Ten years later, people were still dying of cancers and respiratory ailments directly attributable to these biological and chemical agents.

Other symptoms included respiratory, eye and skin problems, severe depression and an increase in suicides, a range of cancers, including leukemias and lymphomas in children, infertility, and a miscarriage rate four times higher than that of other nearby towns. The Halabja genocide was separate from Operation Anfal (1986-89), the campaign conducted by Saddam's regime in order to terrorize the Kurdish rural population.

Dr. Christine Gosden, a professor of medical genetics at the University of Liverpool, on her visit to Halabja ten years after the chemical attack, found evidence of long-term irreversible genetic damage among the survivors. This was due to the effects of the "chemical cocktail Iraqi authorities dumped on an unsuspecting civilian population." Gosden's trip to Iraq, made entirely on humanitarian grounds, showed ten years later effects of chemical attack which included cancers, blindness, and congenital malformations, including leukemias and lymphomas. The gassing of the Halabja citizens was because the Kurdish had sympathized with Iran during the Iran-Iraq war. Gosden said: "What I found was far worse than anything I had suspected, devastating problems occurring ten years after the attack."[7]

Gosden witnessed many Kurds who were covered with horrible skin eruptions. Others went blind and some suffered neurological damage. Couples became infertile and the miscarriage rate was four times higher than that of nearby towns. Some children born during this time had mental retardation, heart defects, harelip, cleft palate, and other major malformations. Half the overall Kurdish population, (approximately 1.5 million) became displaced persons. More than 100,000 were forced into Iran as a result of the chemical attack.

Gosden noted that suicides had become more common and mental illnesses had increased. Although not known at the time, It is now known that poison gas attacks a person's DNA and had similar effects to nuclear radiation. Genetic effects were seen in generations of Kurdish survivors even today. In addition, crops were blighted; domestic animals produced few progeny and those that were born were malformed. However, the viscious attack branded Saddam Hussein and his scientists, including Rasha, as war criminals and became the prime case against them. The atrocity was acknowledged by the world.

The Iraqis blamed the Halabja attack on Iranian forces. At the time of the Halabja attack Iran was reportedly using the blood agent cyanide whereas Iraq was using mustard gas and sarin. However, *Amnesty International (AI), Human Rights Watch,* and *Physicians for Human Rights* disagreed with the Iraqi position since the symptoms they found in exposed patients corresponded to both mustard and sarin gas. There was little evidence to suggest cyanide poisoning used by the Iranians. In contrast, Iraq used multiple chemicals during the attack-mustard gas and nerve agents sarin, tabun, and VX.

Iraq, in its victory, was a wreck. Kurdistan had lost 120,000 citizens by death, 300 were injured and thousands were permanently disabled. Over 60 women and children who had sought shelter in a cave to escape artillery shelling were knowingly incinerated. The Halabja massacre has been recognized by the *Iraqi High Criminal Court* as an act of genocide In March 2010, a decision was welcomed by the Kurdistan Regional Government. The attack was condemned as a crime by the Parliament of Canada.

An investigation into responsibility for the Halabja massacre, by Dr. Jean Pascal Zanders, Project Leader of the *Chemical and Biological Warfare Project at the Stockholm International Peace Research Institute,* concluded in 2007 that Iraq was the sole culprit. The U.S. State Department, however, in the immediate aftermath of the incident, took the official position based on examination of available evidence that Iran was partly to blame.

On December 23, 2005 a Dutch Court sentenced Frans van Anraat, a businessman who bought chemicals on the world market and sold them to Rasha and Saddam Hussein's regime, to 15 years in prison. The Dutch court ruled that Saddam Hussein committed genocide against the people of Halabja. This was the first time a court had described the Halabja attack as an act of genocide. On March 12, March 2008 the government of Iraq announced plans to take further legal action against the suppliers of chemicals use in the poison attack on Halabja.[8] Ali Hassan al-Majid ("Chemical Ali")was condemned to death by hanging by an Iraqi court in January 2010 after being found guilty of orchestrating the Halabja massacre. He was executed on January 25, 2010.

Monument of Halabja

The city of Halabja and its suffering people were virtually forgotten. Then, In March 2003 a controversial monument of Halabja Martyrs was built on the outskirts of Halabja. On March 16, 2006, thousands of Halabja residents rioted at the site, protesting what they perceived as neglect of the living and capitalizing on the tragedy of the Kurdish leadership. The memorial was set on fire; dozens of people were injured and one of the rioters was shot dead by the police.[9] However, it is believed that even today there are mass graves that are undiscovered.

Peter Galbraith Speaks Out

Peter Galbraith, former U.S. ambassador to Croatia from 1993 to 1998, documented Iraqi authorities' attacks against the Kurds in 1988 when he served as senior advisor to the *U.S. Senate Foreign Relations Committee (1979-1993)*. He was one of the first to witness the genocide of the Kurds by Saddam Hussein's Iraqi government during a trip Galbraith made to the region.

At that time, Galbraith wrote: *As we traveled from the Iraqi area to the Kurdish area, we were stunned to see entire villages were gone. These were places that had been inhabited for millennia. The graveyards and mosques were removed. All the wire had been taken down from the electric poles. It had become a desolate region. And we could see where the people had been moved. Iraq called these 'victory cities' but in reality they were a kind of concentration camp.*[10]

Galbraith further reported: *I sat down and dictated, in about an hour, a bill to my secretary. I imposed every sanction I could think of. The legislation banned oil sales, required U.S. to oppose loans, cut off $700 million for agricultural and export credits and banned any export requiring a license. I drafted this and said to myself 'What shall we call this'? It was called The Prevention of Genocide Act of 1988.*[11]

The Prevention of Genocide Act of 1988

The Prevention of Genocide Act of 1988 was a United States bill which addressed Saddam Hussein's weapons of mass destruction attacks on the Kurds. The Iraqi poison attack caused an international outcry against Iraq. U.S. senators Al Gore and Claiborne Pell and Conservative Senator Jesse Helms co-sponsored *The Prevention of Genocide Act of 1988.*

The bill imposed a trade embargo on Iraq only if President Ronald Reagan certified that Iraq was not committing genocide and that Iraq pledged it would not use poison gas again. However, the Reagan administration opposed the bill and eventually the bill became mired in procedural politics. The U.S. House and Senate haggled over the bill. The bill was weaker in the U.S. house. Eventually, Galbraith found himself up against American capitalism.

Agricultural lobbyists took the message into the corridors of the U.S. Congress and warned that the bill would punish Americans who were doing business with Iraq because it would cut off all assistance to

Iraq and stop U.S. imports of Iraqi oil. The bill was defeated, in part due to intense lobbying of Congress by the Reagan-Bush White House and a veto threat from President Ronald Reagan. The President made it known that he was in favor of diplomacy and was acting through the United Nations. The bill died on the final day of the congressional session.

The United States called for the immediate release of 3,000 Americans, and thousands of other Westerners taken hostage, some of whom Saddam Hussein had sent to military and chemical production facilities to serve as human shields in case of an attack. In addition, the Security Council, traditionally paralyzed by ideological and often petty disputes, had unanimously endorsed a financial and trade boycott of Iraq and its oil, 95% of Iraq's foreign exchange. The world was closing in on Saddam Hussein, but not quite yet. In effect, Saddam got away with murder!

Attack on Kuwait: The "Revolution of August 2"

The eruption of hostilities between Iraq and Iran had a terrible effect on Kuwait. After an unsuccessful war with neighbor Iran, the next target for Iraq was Kuwait. Although Kuwait was formally neutral, the Kuwaiti Emir had earlier armed and supported Iraq and Baghdad wholeheartedly, as did Saudi Arabia and all of the Gulf States.

Kuwait was a great source of oil and Saddam Hussein thought it belonged to Iraq, as it had in the past. Together, Kuwait and Saudi Arabia had shipped over 300,000 barrels of crude oil each day for Iraq. But, in 1990 Iraqi Foreign minister Tariq Azziz accused Kuwait of stealing $2.4 billion of oil from Iraq from the Rumaila field which both countries shared. Azziz told an Arab summit meeting in Tunisia: *We are sure Arab states are involved in a conspiracy against us. We want you to know our country will not kneel, our women will not become prostitutes, our children will not be deprived of food.*

Saddam Hussein demanded repayment for the oil which he believed was stolen from the field. In turn, the Kuwaitis viewed Iraq as extorting money from the field. On July 22, 1990 Azziz repeated criticism of Kuwait and on July 26 Saddam Hussein moved an additional 30,000 troops to the Kuwaiti border. By July 28 world leaders believed that Saddam Hussein was only bluffing and would not actually invade Kuwait.

It is hard to imagine two societies more different than Iraq and Kuwait. By July 31 negotiators from Iraq and Kuwait, including President Mubarek of Egypt, attempted to settle the problem, but the meeting collapsed. Although the dispute over the Iraq-Kuwait border had in the past continued for more than 50 years, the dispute over the Iraq-Kuwait border at this time was not anticipated. The U.S. Central Intelligence Agency (CIA) reported that Iraq's potential invasion of Kuwait appeared very probable. "They're ready; they'll go".

Western and Arab analysts disagreed about when and why Saddam Hussein decided to move against Kuwait. Many United States and Middle East experts believed the move was "spontaneous". However, Saddam Hussein kept moving troops to the Kuwaiti border less than a month before the actual invasion. Others viewed Saddam Hussein's actions as "saber-rattling"; trying to bully Kuwait into complying with demands. Others said it was pure ambition on the part of Saddam Hussein, for both money and territory.

Saddam Hussein used economic warfare against Kuwait, demanding $27 billion from Kuwait alone. Iraq owed Kuwait $70 billion, almost half to the Gulf States alone. Saddam Hussein used Kuwait's assets to help repay some of that debt. The Kuwait's pressed for recognizing those borders in exchange for forgiving Iraq's debts.

On August 9, 1990 between 1 and 3 AM, Iraqi forces swept across the Kuwaiti border. It took the Iraqis only five hours to seize and annex tiny Kuwait. The Iraqis employed three waves of troops. First across the border were the Republican Guards, the elite attack forces, battle-hardened veterans of the Iraq-Iran forces; then came the Peoples Army, made up of ragtag, ill-disciplined peasants and thugs, veterans of the Iraq-Iran forces. Finally, there was the regular army and the secret police, who supervised detentions and torture.

Saddam Hussein's invasion of Kuwait was the single largest deployment of American forces overseas since Viet Nam. The United States abruptly sent 100,000 troops to the Middle East. The invasion and seizure of tiny Kuwait, known as the "Revolution of August 2" was trumpeted in the Iraqi State-controlled press. It enraged the world in a way not anticipated, however.

Iraq's harsh invasion of Kuwait came as a shock to not only Kuwait but to the United States and the world, as well. The Bush Administration had announced his decision to send American forces to the area, stressing they were for "purely defensive purposes". President George W. Bush described Saddam Hussein's invasion of Kuwait as a threat to no less "than our way of life", comparing Saddam Hussein's territorial conquests with those of Hitler's. The crisis that unfolded throughout August and September had implications for many in the United States, Europe, and the Middle East that went far beyond what was experienced and the immediate aftermath of Iraq's actions.

When the U.S. allies looked next to win, Saddam Hussein ordered the firing of the Kuwait oil fields, which was an ecological disaster. The U.S. backtracked, isolating-rather than toppling-Saddam Hussein. A cease-fire was declared on February 29, 1991. The attack on Kuwait was considered a blatant violation of international law and convention. Also, Saddam Hussein had sought to change the demographic character of Kuwait, in violation of the *Geneva Conventions.*

The United Nations, at the bequest of the USA, quickly approved five resolutions condemning the invasion, demanding Saddam Hussein's unconditional withdrawal from Kuwait and the immediate release of

3,000 Americans, and thousands of other Westerners taken hostage, some of whom he had sent to military and chemical production facilities to serve as human shields in case of attack.

One such resolution was Resolution 687 (1991). In addition, the Security Council unanimously endorsed a financial and trade boycott of Iraq and its oil, 95% of his country's foreign exchange. The world was closing in on Saddam Hussein and his supporters. Rasha was at the top of Saddam's list of supporters and would remain so until his regime was no more.

The U.N. Security Council, via Resolution 687, said they were conscious also of the statements by Iraq threatening to use weapons in violation of its obligations under the *Geneva Protoco*l for the Prohibition of the Use in War of Asphyxiating, Poisonous or Other Gases, and of Bacteriological Methods of Warfare, signed at Geneva on June 17, 1925. The Council also indicated its knowledge of Iraq's prior use of chemical weapons and affirmed that grave consequences would follow use by Iraq of such weapons.

U.N. Resolution 687 (1991)

Resolution 687 was adopted by the *United Nations Security Council* at its 2981st meeting on April 1991.The Security Council welcomed the restoration to Kuwait of its sovereignty, independence and territorial integrity. It also recognized the return of Kuwait's legitimate Government.

Resolution 687 reads as follows:

The Security Council,

Welcoming the restoration to Kuwait of its sovereignty, independence and territorial integrity and the return of its legitimate Government,

Affirming the commitment of all Member States to the sovereignty, territorial integrity and and political independence of Kuwait and Iraq, and noting the intention expressed by the Member States cooperating with Kuwait under paragraph 2 of resolution 678 (1990) to bring their military presence in Iraq to an end as soon as possible consistent with paragraph 8 of resolution 686 (1991).

Reaffirming the need to be assured of Iraq's peaceful intentions in the light of its unlawful invasion and occupation of Kuwait,

Taking note of the letter sent by the Minister for Foreign Affairs of Iraq on 27 February 1991 and those sent pursuant to resolution 686 (1991),

Noting that Iraq and Kuwait, as independent sovereign States, signed at Baghdad on 4 October 1963 "Agreed Minutes Between the State of Kuwait and the Republic of Iraq Regarding the Restoration of Friendly Relations, Recognition and Related Matters", thereby recognizing formally the boundary between Iraq and Kuwait and the allocation of islands, which were registered with the United Nations in accordance with Article 102 of the Charter of the United Nations and in which Iraq recognized the independence and complete sovereignty of the State of Kuwait within its borders as specified and accepted in the letter of the Prime Minister of Iraq dated 21 July 1932, and as accepted by the Ruler of Kuwait in his letter dated 10 August 1932.

Conscious of the need for demarcation of the said boundary,

Conscious also of the statements by Iraq threatening to use weapons in violation of its obligations under the Geneva Protocol for the Prohibition of the Use in War of Asphyxiating, Poisonous or Other Gases, and of Bacteriological Methods of Warfare, signed at Geneva on 17 June 1925, and of its prior use of chemical weapons and affirming that grave consequences would follow any further use by Iraq of such weapons,

Recalling that Iraq has subscribed to the Declaration adopted by all States participating in the Conference of States Parties to the 1925 Geneva Protocol and Other Interested States, held in Paris from 7 to 11 January 1989, establishing the objective of universal elimination of chemical and biological weapons,

Recalling also that Iraq has signed the Convention on the Prohibition of the Development. Production and Stockpiling of Bacteriological (Biological) and Toxin Weapons and their Destruction, of 10 April, 1972,

Noting the importance of Iraq ratifying this Convention,

Noting moreover the importance of all States adhering to this Convention and encouraging its forthcoming Review Conference to reinforce the authority, efficiency and universal scope of the convention.

Stressing the importance of an early conclusion by the Conference on Disarmament of its work on a Convention on the Universal Prohibition of Chemical Weapons and of universal adherence thereto,

Aware of the use by Iraq of ballistic missiles in unprovoked attacks and therefore of the need to take specific measures in regard to such missiles located in Iraq,

Concerned by the reports in the hands of Member States that Iraq has attempted to acquire materials for a nuclear-weapons program contrary to its obligations under the Treaty on the nonproliferation of Nuclear Weapons of 1 July 1968.

Recalling the objective of the establishment of a nuclear-weapons-free zone in the region of the Middle East.

Conscious of the threat that all weapons of mass destruction pose to peace and security in the area and of the need to work towards the establishment in the Middle East of a zone free of such weapons,

Conscious also of the objective of achieving balanced and comprehensive control or armaments in the region,

Conscious further of the importance of achieving the objectives noted above using all available means, including a dialogue among the States of the region,

Nothing that resolution 686 (1991) marked the lifting of the measures imposed by resolution 661 (1990) in so far as they applied to Kuwait.

Noting that despite the progress being made in fulfilling the obligations of resolution 686 (1991), many Kuwaiti and third country nations are still not accounted for and property remains unreturned,

Recalling the International Convention against the Taking of Hostages, opened for signature at New York on 18 December 1979, which categorizes all acts of taking hostages as manifestations of international terrorism,

Deploring threats made by Iraq during the recent conflict to make use of terrorism against targets outside Iraq and the taking of hostages by Iraq,

Taking note with grave concern of the reports of the Secretary-General of 20 March 1991, and conscious of the necessity to meet urgently the humanitarian needs in Kuwait and Iraq,

Bearing in mind its objective of restoring international peace and security in the area as set out in recent resolutions of the Security Council,

Conscious of the need to take the following measures acting under Chapter VII of the Charter,

1. Affirms all thirteen resolutions noted above, except as expressly changed below to achieve the goals of this resolution, including a formal cease-fire;

A

2. Demands that Iraq and Kuwait respect the inviolability of the international boundary and the allocation of islands set out in the "Agreed Minutes Between the State of Kuwait and the Republic of Iraq regarding the Restoration of Friendly Relations, Recognition and Related Matters", signed by them in the exercise of their sovereignty at Baghdad on 4 October 1963 and registered with the United Nations and published by the United Nations in document 7063, United Nations, Treaty Series, 1964;

3. Calls upon the Secretary-General to lend his assistance to make arrangements with Iraq and Kuwait to demarcate the boundary between Iraq and Kuwait, drawing on appropriate material, including the map transmitted by Security Council document S/22412 and to report back to the security Council within one month;

4. Decides to guarantee the inviolability of the above-mentioned international boundary and to take as appropriate all necessary measures to that end in accordance with the Charter of the United Nations;

5. Requests the Secretary-General, after consulting with Iraq and Kuwait, to submit within three days to the Security Council for its approval a plan for the immediate deployment of a United Nations observer unit to monitor the Khor Abdullah and a demilitarized zone, which is hereby established, extending ten kilometers into Iraq and five kilometers into Kuwait from the boundary referred to in the "Agreed Minutes Between the State

of Kuwait and the Republic of Iraq Regarding the Restoration of Friendly Relations, Recognition and Related Matters" of 4 October 1963; to deter violations of the boundary through its presence in and surveillance of the demilitarized zone;to observe any hostile or potentially hostile action mounted from the territory of one State to the other; and for the Secretary-General to report regularly to the Security Council on the operations of the unit, and immediately if there are serious violations of the zone or potential threats to peace;

6. Notes that as soon as the Secretary-General notifies the Security Council of the completion of the deployment of the United Nations observer unit, the conditions will be established for the Member States cooperating with Kuwait in accordance with resolution 678 (1990) to bring their military presence in Iraq to an end consistent with resolution 686 (1991);

C

7. Invites Iraq to reaffirm unconditionally its obligations under the Geneva Protocol for the Prohibition of the Use in War of Asphyxiating, Poisonous, or Other Gases, and of Bacteriological Methods of Warfare, signed at Geneva on 17 June 1925m and to ratify the Convention on the Prohibition of the Development, Production and Stockpiling of Bacteriological (Biological) and Toxin Weapons and on Their Destruction, of 10 April 1972;

8. Decides that Iraq shall unconditionally accept the destruction, removal, or rendering harmless, under international supervision of:

   a) All chemical and biological weapons and all stocks of agents and all related subsystems and components and all research, development, support and manufacturing facilities;

b) All ballistic missiles with a range greater than 150 kilometers and related major parts, and repair and production facilities;

9. Decides, for the implementation of paragraph 8 above, the following:

a) Iraq shall submit to the Secretary-General, within fifteen days of the adoption of the present resolution, a declaration of the locations, amounts and types of all items specified in paragraph 8 and agree to urgent, on-site inspection as specified below;

b) The Secretary-General, in consultation with the appropriate Governments and, where appropriate, with the Director-General of the World Health Organization, within forty-five days of the passage of the present resolution, shall develop, and submit to the Council for approval, a plan calling for the completion of the following acts within forty-five days of such approval:

(i) The forming of a Special Commission, which shall carry out immediate on-site inspection of Iraq's biological, chemical and missile capabilities, based on Iraq's declarations and the designation of any addition locations by the Special Commission itself;

(ii) The yielding by Iraq of possession to the Special Commission for destruction, removal or rendering harmless, taking into account the requirements of public safety, of all items specified under paragraph 8 (a) above, including items at the additional locations designated by the Special Commission under paragraph 9 (b) (i) above and the destruction by Iraq, under the supervision of the Special Commission, of all its missile capabilities, including launchers, as specified under paragraph 8 (b) above;

(iii) The provision by the Special Commission of the assistance and cooperation to the Director-General of the International Atomic Energy required in paragraphs 12 and 13 below;

10. Decides that Iraq shall unconditionally undertake not to use, develop, construct or acquire any of the items specified in paragraphs 8 and 9 above and requests the Secretary-General, in consultation with the Special Commission, to develop a plan for future ongoing monitoring and verification of Iraq's compliance with this paragraph, to be submitted to the Security Council for approval within one hundred and twenty days of this resolution.

11. Invites Iraq to reaffirm unconditionally its obligations under the Treaty on the nonproliferation of Nuclear Weapons of 1 July 1968;

12. Decides that Iraq shall unconditionally agree not to acquire or develop nuclear weapons or nuclear-weapons-usable material or any subsystems or components or any research, development, support or manufacturing facilities related to the above;to submit to the Secretary-General and the Director-General of the International Atomic Energy Agency within fifteen days of the adoption of the present resolution a declaration of the locations, amounts, and types of all items specified above; to place all of its nuclear-weapons-usable materials under the exclusive control, for custody and removal, of the International Atomic Energy Agency, with the assistance and cooperation of the Special Commission as provided for in the plan of the Secretary-General discussed in paragraph 9 (b) above;to accept in accordance with the arrangements provided for in paragraph 13 below, urgent on-site inspection and the destruction, removal

or rendering harmless as appropriate of all items specified above; and to accept the plan discussed in paragraph 13 below for the future ongoing monitoring and verification of its compliance these undertakings;

13. Requests the Director-General of the International Atomic Energy Agency, through the Secretary-General, with the assistance and cooperation of the Special Commission as provided for in the plan of the Secretary-General in paragraph 9 (b) above, to carry out immediate on-site inspection of Iraq's nuclear capabilities based on Iraq's declarations and the designation of any additional locations by the Special Commission; to develop a plan for submission to the Security Council within forty-five days calling for the destruction, removal, or rendering harmless as appropriate of all items listed in paragraph 12 above; to carry out the plan within forty-five days following approval by the Security Council; and to develop a plan, taking into account the rights and obligations of Iraq under the Treaty on the nonproliferation of Nuclear Weapons of July 1968, for the future ongoing monitoring and verification of Iraq's compliance with paragraph 12 above, including an inventory of all nuclear material in Iraq subject to the Agency's verification and inspections to confirm the Agency safeguards cover al relevant nuclear activities in Iraq, to be submitted to the Security Council for approval within one hundred and twenty days of the passage of the present resolution;

14. Takes note that the actions to be taken by Iraq in paragraphs 8,9,19,11, 12, and 13 of the present resolution represent steps towards the goal of establishing in the Middle East a zone free from weapons of mass destruction and al missiles for their delivery and the objective of a global ban on chemical weapons;

D

15. Requests the Secretary-General to report to the Security Council on the steps take to facilitate the return of all Kuwaiti property seized by Iraq, including a list of any property that Kuwait claims has not been returned or which has not been returned intact;

E

16. Reaffirms that Iraq, without prejudice to the debts and obligations of Iraq arising prior to 2 August 1990, which will be addressed through the normal mechanisms, is liable under international law for any direct loss, damage, including environmental damage and the depletion of natural resources or injury to foreign Governments, nationals and corporations, as a result of Iraq's unlawful invasion and occupation of Kuwait;

17. Decides that all Iraqi statements made since 2 August 1990 repudiating its foreign debt are null and void, and demands that Iraq adhere scrupulously to all of its obligations concerning servicing and repayment of its foreign debt;

18. Decides also to create a fund to pay compensation for claims that fall within paragraph 16 above and to establish a Commission that will administer the fun;

19. Directs the Secretary-General to develop and present to the Security Council for decision, no later than thirty days following the adoption of the present resolution, recommendations for the fund to meet the requirement for the payment of claims established in accordance with paragraph 18 above and for a program to implement the decisions in paragraph s16,17, and 18 above, including:administration of the fund; mechanisms for determining the appropriate level of Iraq' contribution to the fund based on a percentage of the value of the exports of petroleum and petroleum products from Iraq not to exceed a

figure to be suggested to the Council by the Secretary-General, taking into account the requirements of the people of Iraq, Iraq's payment capacity as assessed in conjunction with the international financial institutions taking into consideration external debt service, and the needs of the Iraqi economy; arrangements for ensuring that payments are made to the fund;the process by which funds will be allocated and claims paid; appropriate procedures for evaluating losses, listing claims and verifying their validity and resolving disputed claims in respect of Iraq's liability as specified in paragraph 16 above; and the composition of the Commission designated above;

F

20. Decides, effective immediately, that the prohibitions against the sale or supply to Iraq commodities or products, other than medicine and health supplies, and prohibitions against financial transactions related thereto contained in resolution 661 (1990) shall not apply to foodstuffs notified to the Security Council Committee established by resolution 661 (1990) concerning the situation between Iraq and Kuwait or, with the approval of that Committee, under the simplified and accelerated "no objection" procedure, to materials and supplies for essential civilian needs a identified in the report of the Secretary-general dated 20 March 1991, and in any further findings of humanitarian need by the committee;

21. Decides that the Security Council shall review the provisions of paragraph 20 above every sixty days in the light of the policies and practices of the Government of Iraq, including the implementation of al relevant resolutions of the Security Council, for the purpose of determining whether to reduce or life the prohibitions referred to therein;

22. Decides that upon the approval by the Security Council of the program called for in paragraph 19 above and upon Council

agreement that Iraq has completed all actions contemplated in paragraphs 8,9,10,11, 12, and 13 above, the prohibitions against the import of commodities and products originating in Iraq and the prohibitions against financial transactions related therefore contained in resolution 661 (1991) shall have no further force or effect;

23. Decides that, pending action by the Security Council under paragraph 22 above, the Security Council Committee established by resolution 661 (1990) shall be empowered to approve, when required to assure adequate financial resources on the part of Iraq to carry out the activities under paragraph 20 above, exceptions to the prohibition against the import of commodities and products originating in Iraq;

24. Decides that, in accordance with resolution 661 (1990) and subsequent related resolutions and until a further decision is taken by the Security Council, all States shall continue to prevent the sale or supply, of the promotion of facilitation of such sale or supply, to Iraq by their nations, or from their territories or using their flag vessels or aircraft, of:

    a) Arms and related material of all types, specifically including the sale or transfer through other means of all forms of conventional military equipment, including for paramilitary forces, and spare parts and components and their means of production, for such equipment;

    b) Items specified and defined in paragraphs 8 ad 12 above not otherwise covered above;

    c) Technology under licensing or other transfer arrangements used in the production, utilization or stockpiling of stems specified in subparagraphs (a) and (b) above;

d) Personnel or materials for training or technical support services relating to the design, development, manufacture, use, maintenance or support of items specified in subparagraphs (a) and (b) above;

25. Calls upon all States and international organizations to act strictly in accordance with paragraph 24 above, notwithstanding the existence of any contracts, agreements, licenses or any other arrangements;

26. Requests the Secretary-General, in consultation with appropriate Governments, to develop within sixty days, for the approval of the Security Council, guidelines to facilitate full international implementation of paragraphs 24 and 25 above and paragraph 27 below, and to make them available to all States and to establish a procedure for updating these guidelines periodically;

27. Calls upon all States to maintain such national controls and procedures and to take such other actions consistent with the guidelines to be established by the Security Council under paragraph 26 above as may be necessary to ensure compliance with the terms of paragraph 24 above, and calls upon international organizations to take all appropriate steps to assist in ensuring such compliance;

28. Agrees to review its decisions in paragraphs 22, 23, 24, and 25 above, except for the items specified and defined in paragraph 8 and 12 above, on a regular basis and in any case one hundred and twenty days following passage of the present resolution, taking into account Iraq's compliance with the resolution and general progress towards the control of armaments in the region;

29. Decides that all States, including Iraq, shall take the necessary measures to ensure that no claim shall lie at the instance of the Government of Iraq, or of any person or body in Iraq, or of any

person claiming through or for the benefit of any such person or body, in connection with any contract or other transaction where its performance was affected by reason of the measures taken by the Security Council in resolution 661 (1990) and related resolutions;

G

30. Decides that, in furtherance of its commitment to facilitate the repatriation of all Kuwaiti and third country nationals, Iraq shall extend all necessary cooperation to the International Committee of the Red Cross, providing lists of such persons, facilitating the search by the International Committee of the Red Cross for those Kuwaiti and third country nationals still accounted for;

31. Invites the International Committee of the Red Cross to keep the Secretary-General apprised as appropriate of all activities undertaken in connection with facilitating the repatriation or return of all Kuwaiti and third country nationals or their remains present in Iraq on or after 2 August 1990;

H

32. Requires Iraq to inform the Security Council that it will not commit or support any act of international terrorism allow any organization directed towards commission of such acts to operate within its territory and to condemn unequivocally and renounce all acts, methods and practices of terrorism;

33. Declares that, upon official notification by Iraq to the Secretary-General and to the Security Council of its acceptance of the provisions above, a formal cease-fire is effective between Iraq and Kuwait and the Member States cooperating with Kuwait in accordance with resolution 678 (1990);

34. Decides to remain seized of the matter and to take such further steps as may be required for the implementation of the present resolution and to secure peace and security to the area.

Iraq Underestimates U.S./Coalition

Saddam underestimated the power of the U.S./ Coalition against him. He found Iraq clashed with U.S. laser guided missiles, the casualties reaching up to 100,000 over the course of the war. This great miscalculation cost Saddam Hussein the tiny country of Kuwait, the war, and his power. Not only did Saddam Hussein underestimate the allied forces, but he also underestimated Iraq's defenses.

Weakened considerably, Saddam Hussein had put too much confidence in his weapons systems. At the time, Iraq had little or no air defenses. The U.S. had weakened them even more with well-developed computer viruses. The United States planted a virus in Iraqi computers, disabling much of the already insufficient air defenses. But Saddam Hussein and his team of military officers, including Rasha, had considerable pride. Although the pride kept them going, Saddam Hussein's War Council, including Rasha and her scientists, was failing.

Iraq In Exile

In order to gain further support of the United States, Kuwait offered a permanent air base to the U.S. in the Sheikdom if the ruling Sabbath monarch were restored to power. Other countries pledged their support, including Bahrain, Egypt, and Oman. As a result, Iraq found itself surrounded in a military and economic vise that threatened its very survival.

Iraq was in exile because of its unwarranted action. Not only did Iraq lack natural resources, it was historically vulnerable to invasion from all sides. Baghdad was 70 miles from the Iranian border. Basra was only 13 miles and was under constant siege. Iraq was surrounded To its north was hostile Turkey. The Turks, members of NATO, deeply in debt to the United States and eager to join the ranks of the European Community, were among the first to side with the USA.

To the West was Syria, a fellow Ba'athist state headed by Saddam Hussein's most bitter adversary, Hafez al-Assad. Syria and Iraq had long been divided by ideological and personality conflicts between their leaders and by recent memories of Syria's ardent support for Iran, a non-Arab state, in the Iraq-Iran war. Syria was a hard line foe of Israel and American imperialism. It was also a rich haven for terrorists, At one time Syria had sent troops to help defend Saudi Arabia from Iraqi aggression.

To the west was Jordan, just south of Syria. To the east was Iran, the Shiite fundamentalist rival for hegemony. To neutralize Iran, Saddam Hussein had been forced to relinquish the precious Shoat al-Arab waterway he had won in the war. This had been Saddam Hussein's only access to the Gulf before his seizure of Kuwait. To the south of Iraq was Saudi Arabia. Before the invasion Riyadh had not permitted a single American soldier to be stationed on Saudi soil. But by mid September, the ever cautious Saudis were privately calling upon fellow Arabs and US President George W. Bush to get rid of Saddam Hussein permanently. They were also playing gracious host to more than 200,000 soldiers from more than twenty countries, more than 140,000 of them Americans.

Questions Raised

As a result of the invasion Saddam Hussein faced an awesome US arsenal within striking range of the Saudi border. Present Bush announced his decision on August 8, 1990 to send American forces to the area were "for purely defensive forces". At the time, Bush described Saddam Hussein's invasion of Kuwait as a threat to no less than "our way of life". Further, the President stated: "A half century ago, our nation and the world paid dearly for appeasing and allowing an aggressor who should and could, have been stopped. We are not going to make the same mistake twice".[12]

Other questions were posed at the time: *What was really at stake for the USA in the Persian Gulf? Who was Saddam Hussein? Was Saddam Hussein a modern-day Hitler, the demented irrational "Butcher of Baghdad"? Was Saddam Hussein intent on dominating the Middle East and the West through a stronghold on the supply of oil? Why was almost every government stunned by Iraq's action? How many chemical and biological weapons did Saddam have? Did he have a nuclear bomb? Who was in charge of his bioweapons program? How much authority did Rasha Malikah possess at the time?*

Then U.S. Secretary of State James Baker, the third, asked: *Do we want to live in a world where aggression is made less likely because it is met with a powerful response from the international community, a world where civilized rules of conduct apply, or are we willing to live in a world where aggression can go unchecked, when aggression succeeds because we somehow cannot muster the collective will to challenge it?*[13]

On February 22, 1991, under allied bombs, Saddam Hussein proposed a "withdrawal", but it turned out to be a hoax, a definite mistake. When reviewed by President Bush, it was learned the surrender was much less than acceptable. This was the last straw and the U.S., with its allies, set up to end *Operation Desert Storm*. The U.S. was prepared to end the war with a victory. Miscalculations and mistakes led to Saddam Hussein losing the war, but maintaining his regime.

Meanwhile, *The United Nations Security Council*, at the bequest of the United States, adopted at its 2933rd meeting imposing multilateral sanctions against Iraq. This meant that only supplies intended strictly for medical purposes of food stuffs required in "humanitarian circumstances" would be imported. The decision was left to the Sanctions Committee to determine what constituted "humanitarian circumstances". The USA and its Coalition demanded the immediate release of 3,000 Americans, and thousands of other Westerners who had been take hostage, some of whom Saddam and his War Council (including Rasha) had sent to military and chemical production facilities to serve as human shields in case of attack. The Security Council, traditionally paralyzed by ideological and often petty disputes, unanimously endorsed a financial and trade boycott of Iraq and its oil, 95% of Iraq's foreign exchange. The world was closing in on Saddam and his supporters, including his high-ranking female scientist, Rasha Salih Malikah.

Chapter 9

Economic Sanctions

A half century ago our nation and the world paid dearly for appeasing an aggressor who should, and could, have been stopped. We are not going to make the same mistake twice.

President George H.W. Bush
Comparing Saddam Hussein to Hitler in Remarks
to the Department of Defense Employees

Introduction

Evidence of Iraq's misdeeds had mounted in mid 1989 and by early 1990 pressure was building on Capitol Hill to impose economic sanctions against Saddam Hussein and his regime. At the time, there was no military relationship to speak of so that had become a lesser consideration. When Iraq invaded Kuwait, economic sanctions were applied toward Iraq until March 1991 to pressure Iraq to leave Kuwait. The policy of economic sanctions was also used to pursue political goals, i.e., to remove the Iraqi regime. After that, sanctions were used to get Iraq to comply with the cease-fire terms embodied in U.N. Resolution 687.

Soon after the invasion of Kuwait, President George H.W. Bush announced his decision to send 100,000 American forces to the area, comparing Saddam Hussein to Adolph Hitler when he stated: *"A half century ago our nation and the world paid dearly for appeasing an aggressor who should, and could, have been stopped. We are not going to make the same mistake twice"*.[1] The President's statement is contained in his remarks to the Department of Defense Employees on August 8, 1990. The complete text of his remarks are given below.

Remarks to the Department of Defense Employees

Thank you, Secretary Cheney and General Powell and distinguished members of the Joint Chiefs, General Schwarzkopf, and all of you who do all the work. Thank all of you for joining us today and, really most of all, for all your hard work in defense of freedom and America every day.

Over the past 10 days you've launched what history will judge as one of the most important deployments of allied military power since the Second World War. As I told the American people last week, let no one underestimate our determination to confront aggression. It is you, the men and women of the Department of Defense, who turn these words into deeds that transform hope and promise into reality.

I've just received a wonderful briefing from Secretary Cheney and General Powell and others here at the Pentagon. Our objectives remain clear: the intermediate, complete, and unconditional withdrawal of all Iraqi forces from Kuwait; the restoration of Kuwait's legitimate government; security and stability of Saudi Arabia and the Persian Gulf; and protection of the lives of American citizens abroad. We will achieve these honorable goals.

We've worked for years to develop an international order, a common code and rule of law that promotes cooperation in place of conflict. This order is imperfect; we know that. But without it, peace and freedom are impossible. The rule of law gives way to the law of the jungle. And so, when the question is asked: Where does America stand? I answer: America stands where it always has-against aggression.

Today, the brave American and allied forces are keeping watch along the sands and off the shores of Saudi Arabia. They're here for a purpose: to serve the cause of justice and freedom, a cause the world supports. But Saddam Hussein would have us believe that his unprovoked invasion of a friendly Arab nation is a struggle between Arabs and Americans. And that is clearly false. It is Saddam who lied to his Arab neighbors. It is Saddam who invaded an Arab State. And it is he who now threatens the Arab nation. We, by contrast, seek to assist our Arab friends in their hour of need.

Saddam has claimed that this is a holy war of Arab against infidel-this from the man who has used poison gas against the men, women, and children of his own country; who invaded Iran in a war that cost the lives of more than half a million Moslems; and who now plunders Kuwait. Atrocities have been committed by Saddam's soldiers and henchmen. The reports out of Kuwait tell a sordid tale of brutality.

Saddam would also have us believe that this is a struggle between the haves and the have-nots. But Iraq is one of the haves, for you see, next to Saudi Arabia, Iraq has the largest oil reserves in the world. But thanks to his ruinous policies of war against other Moslems, he-Saddam Hussein-has transferred wealth into poverty. Sadly, it is the Iraqi people who suffer today because of the raw territorial ambition of Saddam Hussein.

Our action in the Gulf is not about religion, greed, or cultural differences, as Iraq's leader would have us believe. What is at stake is truly vital. Our action in the Gulf is about fighting aggression and preserving the sovereignty of nations. It is about keeping our word, our solemn word of honor, and standing by old friends. It is about our own national security interests and ensuring the peace and stability of the entire world. We are also talking about maintaining access to energy resources that are key, not just to the functioning of this country but to the entire world. Our jobs, our way of life, our own freedom, and the freedom of friendly countries around the world would all suffer if control of the world's great oil reserves fell into the hands of that one man, Saddam Hussein.

So, we've made our stand not simply to protect resources of real estate but to protect the freedom of nations. We're making good on long standing assurances to protect and defend our friends who have the courage to stand up to evil and are asking for our help. We are striking a blow for the principle that might does not make right. Kuwait is small. But one conquered nation is one too many.

A half century ago our nation and the world paid dearly for appeasing an aggressor who should and could have been stopped. We're not about to make that same mistake twice. Today Saddam Hussein's Iraq has been cut off by the Arab and Islamic nations that surround it. The Arab League itself has condemned Iraq's aggression. We stand with them, and we are not alone. Sanctions are working. The armies and air forces of Egypt, Morocco, the United Kingdom, and the Gulf Cooperation Council States are shoulder to shoulder with us in Saudi Arabia's defense. Ships of numerous countries are sailing with ours to see that the United Nations sanctions, approved without dissent, are enforced. Together we must ensure that no goods get in and that not one drop of oil gets out.

I am very grateful for the support all of us here are receiving from the American people. The American people are with us. Congress is with us. Our allies are with us. No one should doubt our staying power or our determination. We are in a new era, one full of promise. But events of the past two weeks remind us that there is no substitute for American leadership, and American leadership cannot be effective in the absence of America's strength. I know that this strength does not come cheaply or easily. You pay for it every day in the work you do, in the sacrifices you make, in the time you spend away from your families. I am relying on you to shape the forces of the future, to preserve peace and freedom in the face of new threats and new dangers.

General Powell told me today that it's a great honor, during these dangerous times, to serve as an American soldier. I know its a great honor for me to serve as your Commander in Chief. I thank you. And I join people everywhere in praying for you, for those in the field, and for the United States of America. God bless you all. And thank you for what your doing for your country.

President Bush's announced his decision to send American forces to the area "*for purely defensive forces*" .It was the single largest deployment of American forces sent overseas since the Viet Nam War. Not surprisingly, at the time, some Iraqi leaders asserted there was a strange conspiracy by Kuwait and the United States. to destroy Baghdad. In effect, the United Nations imposed a blanket ban on all imports and exports in Iraq. Iraq was cut off from the world under a siege known as economic sanctions. Once the decision was made-economic sanctions were imposed on Iraq.

Initially, the sanctions were viewed as a short-term penalty to induce nationwide poverty and to force Iraq to withdraw from Kuwait, as well as to bring down the regime of Saddam Hussein. The United Nations Security Council pulled together and endorsed a financial boycott of Iraq and its oil and most of Iraq's foreign exchange. At its 293rd meeting held on August 6, 1990, *Resolution 661* was adopted by the Security Council, imposing comprehensive sanctions on Iraq following Iraq's invasion of Kuwait.

U.N. Resolution 661 (1990)[2]

Resolution 661 read as follows:

The Security Council,

Reaffirming its resolution 660 (1990) of 2 August 1990,

Deeply concerned that the resolution has not been implemented and that the invasion by Iraq on Kuwait continues with further loss of human life and material destruction,

Determined to bring the invasion and occupation of Kuwait to an end and to restore the sovereignty, independence and territorial integrity of Kuwait,

Noting that the legitimate Government of Kuwait has expressed its readiness to comply with resolution 660 (1990),

Mindful of its responsibilities under the Charter of the United Nations for the maintenance of international peace and security,

Affirming the inherent right of individual or collective self-defense, in response to the armed attack by Iraq against Kuwait, in accordance with Article 51 of the Charter,

Acting under Chapter VII of the Charter of the United Nations,

1. Determines that Iraq so far has failed to comply with paragraph 2 of resolution 660 (1990) and has usurped the authority of the legitimate Government of Kuwait;

2. Decides, as a consequence, to take the following measures to secure measures to secure compliance of Iraq with paragraph 2 of resolution 660 (1990) and to restore the authority of the legitimate Government of Kuwait;

3. Decides that all States shall prevent:

   a) The import into their territories of all commodities and products originating in Iraq or Kuwait t exported therefrom after the date of the present resolution;

b) Any activities by their nationals or in their territories which would promote or are calculated to promote the export or transshipment of any commodities or products from Iraq or Kuwait; and any dealings by their nationals or their flag vessels or in their territories in any commodities or products originating in Iraq or Kuwait and exported therefrom after the date of the present resolution, excusing in particular any transfer of funds to Iraq or Kuwait for the purposes of such activities or dealings;

c) The sale or supply by their nationals or from their territories or using their flag vessels of any commodities or products, including weapons or any other military equipment, whether or not originating in their territories but not including supplies intended strictly for medical purposes, and, in humanitarian circumstances, foodstuffs, to any person or body in Iraq or Kuwait or to any person or body for the purposes of any business carried on or operated from Iraq or Kuwait, and any activities by their nationals or in their territories which promote or are calculated to promote such sale or supply of such commodities or products;

4. Decides that all states shall not make available to the Government of Iraq or to any commercial, industrial or public utility undertaking in Iraq or Kuwait, any funds or any other financial or economic resources and shall prevent their nationals and any persons within their territories from receiving from their territories or otherwise making available to that Government or to any such undertaking any such funds or resources and from remitting any other funds to persons or bodies within Iraq or Kuwait, except payments exclusively for strictly medical or humanitarian purposes and, in humanitarian circumstances, foodstuffs;

5. Calls upon all States, including States nonmembers of the United Nations, to act strictly in accordance with the provisions

of the present resolution notwithstanding any contract entered into or license granted before the date of the present resolution;

6. Decides to establish, in accordance with rule 28 of the provisional rules of procedure of the Security council, a Committee of the Security Council consisting of all the members of the Council, to undertake the following tasks and to report on its work to the Council with its observations and recommendations:

    a) To examine the reports on the progress of the implementation of the present resolution which will be submitted by the Security-General;

    b) To seek from all States further information regarding the action taken by them concerning the effective implementation of the provisions laid down in the present resolution;

7. Calls upon all States to cooperate fully with the Committee in the fulfillment of its task, including supplying such information as may be sought by the Committee in pursuance of the present resolution;

8. Requests the Secretary-General to provide all necessary assistance to the Committee and to make the necessary arrangements in the Secretariat for the purpose;

9. Decides that, notwithstanding paragraphs 4 through 8 above, nothing in the present resolution shall prohibit assistance to the legitimate Government of Kuwait, and calls upon all States:

    a) to take appropriate measures to protect assets of the legitimate Government of Kuwait and its agencies;

    b) Not to recognize any regime set up by the occupying Power;

10. Requests the Secretary-General to report to the Council on the progress of the implementation of the present resolution, the first report to be submitted within thirty days;

11. Decides to keep this item on its agenda and to continue its efforts to put an early end to the invasion by Iraq.

ht.://www.fas.org/news/un/iraq/sees/sres0661.htm

Effects of the Sanctions

The immediate consequences of eight years of sanctions meant a dramatic fall in living standards for the Iraqi people, collapse of the infrastructure, and a serious decline in the availability of public services. Long term damage included heightened levels of crime, corruption, and violence. Before the Gulf War, living standards in Iraq were approaching those of Southern Europe, including free education, ample electricity, modern farming, a large middle class, and according to the World Health Organization (WHO), access to health care for 93 percent of the population.[3]

The effects of the blockade on Iraq, which at one time imported 70 per cent of its food, was that its entire infrastructure was reduced to rubble by possibly the most intense campaign in history. Sanctions led to a steep increase in hunger, disease and death throughout the entire Iraqi society. Later, United States inspectors agreed that Iraq had abandoned its Weapons of Mass Destruction programs, but asserted Iraq had an intention to pursue these programs once United Nations sanctions were ever lifted.[4]

From the early 1950's until the major cessation of exports in 1990 the Iraqi economy had been dominated by the oil sector[5] Prior to the 1991 Persian Gulf War, Iraq had one of the highest per-capita food availability ratings in the region, due to its relative prosperity and capacity to import large quantities of food, which met up to two-thirds of food requirements. The imposition of UN sanctions in August 1990 significantly constrained Iraq's ability to earn the foreign currency needed to import sufficient quantities of food to meets its needs. As a consequence, food shortages and malnutrition became progressively more severe and chronic.[6]

The Iraqi government immediately recognized that these international actions could ultimately destroy its ability to survive and ultimately Iraq withdrew from Kuwait. In spite of the withdrawal, however, the United Nations Security Council (UNSC) decided to maintain sanctions in Iraq. In 1990 Central Intelligence Agency (CIA) William Webster spoke to the U.S. Congress and reported: "*Our judgment has been and continues to be that there is no assurance or guarantee that economic hardships will compel Saddam to change his policies or lead to internal unrest that would threaten his regime*".[7]

Once sanctions were in place, average food intake in Iraq declined by one third; growth stunting and wasting were as common as in the worst Third World countries and mortality for children under five years old was almost two and one half times the pre-sanction rate. At the time, It was estimated that 5,000-7,000 or more children under five died each month as a result of the sanctions. In the late 1980's, the mortality for Iraqi children under five was 50 per thousand. By 1999, the number had reached 130 per thousand. In 1998 500,000 Iraqi children had died due to economic sanctions-related effects, largely dysentery and acute diarrhea.[8]

On the basis of the downward trends in mortality rates which were observed between 1960 and 1990, it was estimated that some 500,000 excess deaths of children under the age of five may have occurred during the period from 1991-1998.[9,10] As further evidence of the imposed sanctions, Iraqi hospitals were full of children dying from diseases, especially Kwashiorkor, marasmus (diseases of severe malnutrition) and cancer.[11]

During this time, physicians, nurses, and other medical personnel were forced to stand by and helplessly watch because there was no medicine while thousands died before their eyes. Dismayed at the effects of the sanctions on the Iraqi people, Rasha wrote:*"Iraqi death rates have increased significantly, with cancer representing a significant cause of mortality, especially in the south and among children"*. For her work, Rasha earned international recognition for her research and documentation of the rise of cancers and other diseases among Iraqi children and war veterans due to economic sanctions.[12,13]

Other economic setbacks became evident because of additional shortages of equipment, machinery, and essential medicines needed for humans and animals. Food safety also became a major problem. In the first half of the 1990's economic decline proceeded very rapidly and major setbacks became known. Prior to the Gulf War there had been an improving standard of living for the vast Iraqi population, including provision of health care. Infant mortality had declined about 120 per 1,000 live births in 1960 to about 45 per 1,000 by late 1980's.[14] However, following imposition of sanctions in 1990, the infant mortality returned to more than 100 per 1,000 by 1998.[15]

By early March 1991 the U.S. and its Coalition had driven Iraqi forces completely out of Kuwait. In the process, however, Iraq's civilian infrastructure, including factories, water and sewage plants, and oil sales were almost depleted. In addition, electric power stations were disabled. Water and sanitation systems were critical, with the Basra area the most serious. As a result, there were cases of infectious diseases, especially typhoid fever, hepatitis, and gastroenteritis in the hot summer months.[16]

The Food and Agricultural Organization (FAO), a United Nations agency in Baghdad, was unequivocal about returning Iraq to prosperity. Questions as to the decline surfaced among the FAO and the Iraqi people. It had the water; it had the land; it had the expertise. Yet, production had declined. because there were no pesticides, no fertilizers, no improved seeds. The grain crises became severe. The animal husbandry and poultry situation were as severe as the grain crises. The Iraqi government soon discovered that without vaccines and other medicines, animals, like humans and plants, could not survive.16

According to a 1995 FAO report, dairy herds were down forty percent since 1990. This number fell from 1,512,000 to only one million by 1995. Water buffalo and goat herds declined even more. All this happened because necessary imports were not available. It was a kind of agricultural sabotage that ensured Iraq could not become food self sufficient, but rather depended on the United States and other nations for its substinence.

The sanctions had a distinct effect on the lives of Iraqi people. UNICEF published an independent report by a consult, Eric Hoskins,[17] on the impact of sanctions and UNICEF's perspective. They found multiple impacts caused by the sanctions.

Hoskins divided his findings into **Direct Effects (immediate)**; **Short-term Effects (intermediate)**; and, **Long-term Effects (chronic.)**

Among the **direct effects (<u>immediate</u>)** were decreased Imports. These included medicines, food imports, agricultural Inputs (fertilizer, pesticides, spare parts, industrial/commercial inputs/parts, and other spare parts, fuel, educational materials, water, purification/supply, and inputs. Decreased Exports included impact on export earnings, access to foreign currency, etc. Decrease in Communications included telecommunications, media. There was also a significant impact on diplomatic efforts.

Among the **short term effects (<u>intermediate</u>) effects** were: health which included a deterioration in health status; increased morbidity and mortality (especially child, maternal and prenatal mortality, low birth-weight babies, infectious diseases, epidemics, malnutrition; deterioration in water quantity and quality; deterioration in health services; decrease in available medicines, vaccines, laboratory and diagnostic tests; breakdown of medical, x-ray, and lab equipment's;. food security, higher market prices for basic foodstuffs; "entitlement" problems in obtaining food; shortages of basic food items; decrease in household diet and caloric intake; decreased agricultural and production; decrease in livestock production; black market purchases. Economics: was also event and included decreased export earnings; decreased trade leading to closure of business and industry; inflation; unemployment; emergence of black market; decreased wages, purchasing power; increase in personal/household loans; decreased economic activity (industry, commerce, agriculture, etc) due to lack of trading partners, resources, funds, inputs.

Among the **Long Term Effects (chronic) effects** were: reduction in the overall (general) health status of the population;deterioration in health services and diminished national capacity to provide care; loss of previous gains in preventive and curative care services; resurgence of illness and disease associated with poverty (e.g. epidemics, infectious disease). **Economic effects** included chronically decreased economic activity and a decline in revenue from all sources. Also, there were significant declines in the GDP, GNP, per capital income; loss of trade partners, regional /international trade interests; chronically high unemployment; collapse of public and private infrastructure; decline in public education. **Social effects** included increased poverty; increases in social inequality (income gap between rich and poor); social upheaval, violence distress; decreased in social cohesion; psychoscial impact difficult to measure. Political effects included an impact on democracy; impact on human rights, previously-observed; democratic freedoms; change in regional balance of power, security.

Double-Embargo

The Iraqi government also withdrew funding and services for three northern governorates and imposed it's own economic blockage on the region in October 1991, leading to the creation of a *de-facto*

Kurdish-controlled region known as Iraqi Kurdistan. However, the international community did not alter the scope of its sanctions on Iraq, which remained in force over the whole of Iraq.

This "double embargo' (sanctions) imposed by the international community and also by Saddam Hussein and his government of Iraq encouraged the development of a nonproductive economy based on revenues derived from custom duties, as well as smuggling to Turkey, Iran, and government-controlled areas of Iraq. This resulted in four years of internal fighting (1993-1997) which eventually resulted in the virtual collapse of the Kurdish Regional Administration.

Rasha's Criticisms of the Sanctions

Rasha Malikah was an Iraqi patriot dedicated to staying and helping her people get through the hideous ordeal of the many economic sanctions they faced. In her work as both a scientist and an educator, she was handicapped by a lack of books and scientific supplies. She said: "We cannot plan. No one can plan beyond the next day. It is hard for my students to imagine a career and move toward a goal. We proceed from day to day, just as I do with my own children". Later, as the Iraqis struggled with the sanctions, Rasha replied: *It is difficult for anyone to have a plan. You have a plan when you have a settled situation-known circumstances. We don't have that anywhere in Iraq. My immediate plan is to provide tomorrows means for life for my children, to help my students into another successful day. After that, I don't know.*[18]

Rasha and her team of scientists were angered, concerned, and deeply critical of the sanctions imposed upon Iraqi citizens. The entire area and population were affected. Iraq was besieged with diseases. Waterborne parasites and bacteria, plus pronounced malnutrition that existed, were responsible for many different diseases, and for wasting and death of Iraqi citizens. In addition, there was a sharp increase in spontaneous abortions, cancers, and other "new diseases", Rasha was also concerned about the combined effects of pollution and radiation on the entire Iraqi population which rendered them even more susceptible to these conditions.

Deeply critical of the sanctions imposed against Iraq, Rasha believed that the sanctions constituted the most comprehensive economic blockage of any country in modern times and was, in effect, a war against civilians that preferentially targeted children, the elderly, and the poor. She cited as evidence severe malnutrition, lack of medical care, and increased prevalence of diseases. These were compounded by destroyed electricity and water sanitation plus an infrastructure caused by bombing, high levels of malnutrition and increased mortality which were inevitable. It was estimated that the number killed by economic sanctions alone was more than one million. Half ot the dead were children under the age of five.

Failure of the United States and the United Nations to explicitly spell out what was needed for the sanctions to be lifted led to Iraq suspending its cooperation with United Nations inspectors in December of 1988.

"Oil for Food Programme"[19]

The *Oil for Food Program*me was started in 1996 as the effects of sanctions on the Iraqi people were getting worse. The idea was that Iraq would be able to sell oil and use 60% of that money to purchase humanitarian goods. The remainder of that money went to things like reparations to Kuwait, coalition operations in Kuwait, etc. Some goods took a long time to be approved, and it appeared that a lot of money was raised.

The *Oil for Food Programme* was intended as a limited and temporary emergency measure. It was first offered to Iraq in 1991 and at that time was rejected. It was first put into place in 1996.In the program, Iraq was allowed to sell a limited amount of oil until 1999. After that, limits would be removed.

Unfortunately, the program was rife with corruption through smuggling and kickbacks. Saddam Hussein managed to acquire more than $10 billion in illicit funds. Others claimed Saddam's access to illicit funds were over $ 21.5 billion, money which went to terrorists around the world, plus post-Saddam insurgency.

The mainstream media often made the claim that the United Nations allowed Saddam Hussein to steal billions of dollars from oil sales. While Iraq was under UN sanctions (1991-2003) is when the Hussein regime obtained most of its illicit funds.

The "Oil for Food Programme" was highly transparent. A common theme in the criticism for the program lacked accountability. or oversight. However, the program was highly transparent. Julian Borgen and Jania Wilson reported for the British newspaper, *The Guardian*, that United States oil purchases accounted for 52% of the kickbacks paid to the Iraqi regime in return for sales of cheap oil. This was more than the rest of the world put together.

The Bush administration was made aware of illegal sales and kickbacks paid to the Saddam regime but according to Borgen and Wilson did nothing to stop them. In a few cases, they actually facilitated the illicit oil sales. Further, the U.S. Treasury apparently played delaying and avoiding tactics to repeated requests by United Nations staff and the US State department for information on transactions on a United States oil company. The U.S. military allowed some oil to be smuggled out of Iraq, even assuring a U.S. oil company that the oil they had illegally purchased would not be confiscated.

On April 14, 1995, the UN Security Council adopted Resolution 986, the "Oil for-Food Programme", which permitted Iraqi oil sales to finance imports of food and other essential humanitarian supplies.

Resolution 986 (1995)[20]

Resolution 986 (1995) reads as follows:

The Security Council,

Recalling its previous relevant resolutions,

Concerned by the serious nutritional and health situation of the Iraqi population, and by the risk of a further deterioration in this situation,

Convinced of the need as a temporary measure to provide for the humanitarian needs of the Iraqi people until the fulfillment by Iraq of the relevant Security Council resolutions, including notably resolution 687 (1991) of 3 April 1991, allows the Council to take further action with regards to the prohibitions referred to in resolution 661 (1990) of August 1990, in accordance with the provisions of those resolutions,

Convinced also of the need for equitable distribution of humanitarian relief to all segments of the Iraqi population throughout the country,

Reaffirming the commitment of all Member States to the sovereignty and territorial integrity of Iraq,

Acting under Chapter VII of the Charter of the United Nations,

1. *Authorizes* States, notwithstanding the provisions of paragraphs 3 (a), 3 (b) and 4 of Resolution 661 (1990) and subsequent relevant resolutions, to permit the import of petroleum an petroleum products originating in Iraq, including financial and other essential transactions directly relation thereto, sufficient to produce a sum not exceeding a total of one billion United States dollars ever 90 days for the purposes set out in this resolution and subject to the following conditions:

a) Approval by the Committee established by resolution 661 (1990), in order to ensure the transparency of each transaction and its conformity with the other provisions of this resolution, after submission of an application by the State concerned, endorsed by the Government of Iraq, for each proposed purchase of Iraqi petroleum and petroleum products, including details of the purchase price at fair market value, the export route, the opening of a letter of credit payable t the escrow account be established by the Secretary-General for the purpose of this resolution, and of any other directly related financial or other essential transaction;

b) Payment of the full amount of each purchase of Iraqi petroleum and petroleum products directly by the purchaser in the State concerned into the escrow account to be established by the Secretary-General for the purposes of this resolution;

2. *Authorizes* Turkey, notwithstanding the provisions of paragraphs 3 (a) and 4 of resolution 661 (1990) and the provisions of paragraph 1 above, to permit the import of petroleum and petroleum products originating in Iraq sufficient, after the deduction of the percentage referred to in paragraph 8 (c) below for the Compensation Fund, to meet the pipeline tariff charges, verified as reasonable by the independent inspection agents referred to in paragraph 6 below, for the transport of Iraqi petroleum and petroleum products through the Kirkuk-Yumurtalik pipeline in Turkey authorized by paragraph 1 above;

3. *Decides* that paragraphs 1 and 2 of this resolution shall come into force at 00.1 Eastern Standard Time on the day after the President of Council has informed the members of the Council that he has received the report from the Secretary-General requested in paragraph 13 below, and shall remain in force for

an initial period of 180 days unless the Council takes other relevant action with regard to the provisions of resolution 661 (1990);

4. *Further decides* to conduct a thorough review of all aspects of the implementation of this resolution 90 days after the entry into force of paragraph 1 above and again prior to the end of the initial 180 day period, on receipt of the reports referred to in paragraphs 11 and 12 below, and *expresses its intention,* prior to the end of the 180 day period, to consider favorably renewal of the provisions of this resolution, provided that the reports referred to in paragraphs 11 and 12 below indicate that those provisions are being satisfactorily implemented;

5. *Further decides* hat the remaining paragraphs of this resolution shall come into force forthwith;

6. *Directs* the Committee established by resolution 661 to monitor the sale of petroleum and petroleum products to be exported by Iraq via the Kirkuk-Yumurtalik pipeline from Iraq to Turkey and from the Mina al-Bakr oil terminal, with the assistance of independent inspection agents appointed by the Secretary-General, who will keep the Committee informed of the amount of petroleum and petroleum products exported from Iraq after the date of entry into force of paragraph 1 of this resolution, and will verify that the purchase price of the petroleum and petroleum products is reasonable in the light of prevailing market conditions, and that, for the purposes of the arrangements set out in this resolution, the larger share of the petroleum and petroleum products is shipped via the Kirkuk-Yumurtalik pipeline and the reminder is exported from the Min al-Bakr oil terminal;

7. *Requests* the Secretary-General to establish an escrow account for the purposes of this resolution, to appoint independent

and certified public accountants to audit, and to keep the Government of Iraq fully informed;

8. *Decides* that the funds in the escrow account shall be used to meet the humanitarian needs of the Iraqi population and for the their purposes, and *requests* the Secretary-General to use the funds deposited in the escrow account:

   a) To finance the export to Iraq, in accordance with the procedures of the Committee established by resolution 661 (1990), of medicine, health supplies, foodstuffs, and materials and supplies for essential civilian needs, as referred to in paragraph 20 of resolution 687(1991)provided that

      (i) Each export of goods is at the request of the Government of Iraq;

      (ii) Iraq effectively guarantees their equitable distribution, on the basis of a plan submitted to and approved by the Secretary-General, including a description of the goods to be purchased;

      (iii) The Secretary -General receives authenticated confirmation that the exported goods concerned have arrived in Iraq;

   b) To complement, in view of the exceptional circumstances prevailing in the three Governments mentioned below, the distribution by the Government of Iraq of goods imported under this resolution, in order to ensure an equitably distribution of humanitarian relief to all segments of the Iraqi population throughout the country, by providing between 130 million and 150 million United States dollars every 90 days to the *United Nations interagency Humanitarian Program* operating within the sovereign territory of Iraq in the three northern Governments of Dihouk, Arbil and Suleimanlyeh, except that if less than one billion United States dollars worth of Petroleum products is sold during any 90 day period, the Secretary-General may provide a proportionately smaller amount for this purpose;

c) To transfer to the Compensation Fund the same percentage of the funds deposited in the escrow account as that decided by the Council in paragraph 2 of resolution 705 (1991) of 15 August 1991;

d) To meet the costs to the United Nations of the independent inspection agents and the certified public accountants and the activities associated with implementation of this resolution;

e) To meet the current operating costs of the Special Commission, pending subsequent payment in full of the costs of carrying out the tasks authorized by section C of resolution 687 (1991);

f) To meet any reasonable expenses, other than expenses payable in Iraq, which are determined by the Committee established by resolution 661 (1990) to be directly related to the export by Iraq of petroleum and petroleum products permitted under paragraph 1 above or to the export to Iraq, and activities directly necessary therefor, of the parts and equipment permitted under paragraph 9 below;

g) To make available up to 10 million United States dollars every 90 days from the funds deposited in the escrow account for the payments envisaged under paragraph 6 of resolution 778 (1992) of 2 October 1992;

9. *Authorizes* States to permit, notwithstanding the provisions of paragraph 3 (c) of resolution 661 (1990):

    a) The export to Iraq of the parts and equipment which are essential for the safe operation of the Kirkuk-Yumurtalik pipeline system in Iraq, subject to the prior approval by the Committee established by resolution 661 (1990) of each export contract;

    b) Activities directly necessary for the exports authorized under subparagraph (a) above, including financial transactions related thereto;

10. *Decides* that, since the costs of the exports and activities authorized under paragraph 9 above are precluded by paragraph 4 of resolution 661 (1990) and by paragraph 11 of resolution 778 (1991) from being met from funds frozen in accordance with those provisions, the cost of such exports and activities may, until funds begin to be paid into the escrow account established for the purposes of this resolution, and following approval in each case by the Committee established by resolution 661 (1990), exceptionally be financed by letters of credit, drawn against future oil sales the proceeds of which are to be deposited in the escrow account;

11. *Requests* the Secretary-General to report to the Council 90 days after the date of entry into force of paragraph 1 above and again prior to the end of the initial 180 day period, on the basis of observation by United Nations personnel in Iraq, and on the basis of consultations with the Government of Iraq, on whether Iraq has ensured the equitable distribution of medicine, health supplies, foodstuffs, and materials and supplies for essential civilian needs, financed in accordance with paragraph 8(a(above, including in his reports any observations he may have on the adequacy of the revenues to meet Iraq;s humanitarian

needs, and on Iraq's capacity to export sufficient quantities of petroleum and petroleum products to produce the sum referred to in paragraph 1 above;

12. *Requests* the Committee established by resolution 661 (1990), in close coordination with the Secretary-General, to develop expedited procedures as necessary to implement the arrangements of paragraphs 1,2,6,8,9,and 10 of this resolution and to report to the Council 90 days after the date of entry into force of paragraph 1 above and again prior to the end of the initial 180 day period on the implementation of those arrangements;

13. *Requests* the secretary-general to take the actions necessary to ensure the effective implementation of this resolution, authorizes him to enter into any necessary arrangements or agreements, and requests him to report to the Council when he has done so;

14. Decides that petroleum and petroleum products subject to this resolution shall while under Iraqi title be immune from legal proceedings and not be subject to any form of attachment, garnishment or execution, and that all States shall take any steps that may be necessary under their respective domestic legal systems to assure this protection, and to ensue that the proceeds of the sale are not diverted from the purposes laid down in this resolution;

15. *Affirms* that the escrow account established for the purposes of tis resolution enjoys the privileges and immunities of the United Nations;

16. *Affirms* that all persons appointed by the Secretary-General for the purpose of implementing this resolution enjoy privileges and immunities as experts on mission for the United Nations in accordance with the Convention on the Privileges and

Immunities of the United Nations, and *requires* the Government of Iraq to all them full freedom of movement and all necessary facilities for the discharge of their duties in the implementation of this resolution;

17. Affirms that nothing in this resolution affects Iraq's duty scrupulously to adhere to all of its obligations concerning servicing and repayment of its foreign debt, in accordance with the appropriate international mechanisms;

18. Also *affirms* that nothing in this resolution should be construed as infringing the sovereignty or territorial integrity of Iraq;

19. *Decides* to remain seized of the matter.

The implementation of *Resolution 986* undoubtedly eased the food-supply situation in Iraq but did not end the overall civilian suffering in Iraq. Severe limitations in supplies and medical services and nutritional problems in women, children, and the elderly continued to be serious and widespread.

Severe forms of malnutrition were evidenced by clinical signs such as edema, and severe wasting in ribs, face, and limbs of the victims. Serious vitamin (particularly vitamin A) and mineral deficiencies were evident, and anemia was very common. Adults under 25 who were malnourished were markedly underweight for their age. The number killed by economic sanctions alone in 1990 was more than one million, according to estimates by different UN agencies.

A Decade Later

After one decade of feeling the full effects of economic sanctions, depleted uranium, electromagnetic pollution and chemical pollution, cancers in Iraq had risen up to tenfold. In Basra, radiation levels in flora and fauna reached 84 times the World Health Organization's (WHO) recommended safe limit. Physicians at Basra General Hospital in southern Iraq, kept a photographic record of all babies born with no ears or eyes, foreshortened limbs, and abnormal genitalia. Also seen were babies with internal organs on the outside, and grotesquely deformed heads and bodies.[21]

John Pilger stated: *"The change in 10 years is unparalleled, in my experience"*,

Anupama Rao Singh, UNICEF's senior representative in Iraq told me that. In 1989, the literacy rate in Iraq was 95%; and 93% of the population had free access to modern health facilities. Parents were fined for failing to send their children to school. The phenomenon of street children or children begging on the streets was unheard of. Iraq had reached the stage where basic indicators were used to measure the overall well-being of human beings, including children, were some of the best in the world. Once at the bottom 20%,in 10 years child mortality had gone from one of the lowest in the world, to the highest'.[22]

Other Iraqi physicians observed a record increase in the number of other children born deformed, with missing limbs, eyes and ears. Also seen were cleft palate, club foot, enlarged heads; some without formed genitalia. Many malformed babies were born at the time. Increased spontaneous abortions were prominent. UNICEF confirmed that an estimated 500,000 or more children died in Iraq in the decade following the Gulf War than had died in the previous decade.

Dr. Neboysha Ljepovic, professor of physics at London University, estimated that *"external and internal exposures of Iraqi and non-Iraqi Gulf War veterans could be 3,000 to 21,000 additional cancers for every 100,000 veterans"*.[23] In Mosul in northern Iraq, studies undertaken by four universities showed a fivefold increase in cancers after 1991. An informal survey in the area counted 20 malformed babies in 160 houses; the majority had fathers who had served in the Gulf War.

U.S. Journalist Barbara Aziz reported a farmer remarking that there were few marriages as a result, saying, *Young people fear the birth of malformed fetuses and still births. We look around our village, everyone knows couples who have deformed babies.*[24] Soaring cancers in Iraq since the Gulf War are well documented. In a Johns Hopkins University paper presented by Professor Mikdem M Falak in Baghdad in December. It was estimated that if the number of cancers continued on the present upward curve, 44 per cent of the Iraqi population would be expected to develop cancer within the next ten years following the war.[25]

Rasha reported that after the Persian Gulf War air pollutants, measured as total suspended particles, increased in Baghdad by 705 per cent. Eight years after the war ended, the values were 887 percent of the levels recommended by the World Health Organization (WHO). Concentrations of heavy metals such as lead and cadmium also increased and were responsible for various health problems, genetic defects, cardiovascular damage, and numerous kinds of cancers. Infectious diseases were also increased due to invasion of microbes, insects, and rodents.

Rasha also drew attention to the illegality in international law of the use of weapons whose effects continued to kill and pollute long after the war was over. DU weapons remain toxic and radioactive for four-and-a-half billion years.[26] Some 150 peace organizations and non-governmental organizations (NGOs) around the world protested the Resolution. Yet Resolution 986 was passed by the 150 groups by what they described as bribes and threats by the USA on other members of the Council. It was described as political legitimacy to US rules in Iraq.

On May 22, 2003 the United Nations Security Council voted to lift the sanctions against Iraq, Saddam's regime having been toppled. The vote was 14 to 1 (Syria refused to vote) . The passing of the Resolution was controversial. In the past, the United States and the United Kingdom had been most vocal in maintaining sanctions against Iraq. Now, they were the most vocal in lifting them. The Resolution did not specify the role of the International Atomic Energy Agency (IAEA) in declaring Iraq free of weapons of mass destruction. It did not end the UN arms embargo against the country and it did not clarify the UN's role in a future Iraq.

Chapter 10

The 2003 War

The enemies you confront will come to know your skill and bravery. The people you liberate will witness the honorable and decent spirit of the American military.

President George W. Bush
Address to the Nation, March 19, 2003

Operation Iraqi Freedom

The media played a significant role in the Weapons of Mass Destruction (WMD) controversy. In the buildup to the 2003 War the *New York Times* published a number of stories claiming to prove that Iraq possessed WMD. One story, written by Judith Miller, help persuade the American public that Iraq had WMD. Miller's sources were introduced to her by Ahmed Chalabi, an Iraqi exile. Her stories were followed up with television appearances by Colin Powell, Donald Rumsfeld, and Condoleeza Rice, all pointing to the story as part of the basis for taking military action against Iraq.

The U.S. and Britain, along with other allies, launched **Operation Iraqi Freedom** on March 19, 2003. This was a military campaign that quickly brought the end of Saddam Hussein's regime and ultimately resulted in his capture.

On March 19, 2003 President George W. Bush addressed the nation from the Oval Office at 10:16 EST:[1]

THE PRESIDENT: My fellow citizens, at this hour, American and coalition forces are in the early stages of military operations to disarm Iraq, to free it's people and to defend the world from grave danger.

On my orders, coalition forces have begun striking selected targets of military importance to undermine Saddam Hussein's ability to wage war. These are opening stages of what will be a broad and concerted campaign. More than 35 countries are giving crucial support-from the use of naval and air bases, to help with intelligence and logistics, to the deployment of combat units. Every nation in this coalition has chosen to bear the duty and share the honor of serving in our common defense.

To all the men and women of the United States Armed Forces now in the Middle East, the peace of a troubled world and the hopes of an oppressed people now depend on you. That trust is well placed.

The enemies you confront will come to know your skill and bravery. The people you liberate will witness the honorable and decent spirit of the American military. In this conflict, America faces an enemy who has no regard for conventions of war or rules of morality. Saddam Hussein has placed Iraqi troops and equipment in civilian areas, attempting to use innocent men, women and children as shields for his own military-a final atrocity against his people.

I want Americans and all the world to know that coalition forces will make every effort to spare innocent civilians from harm. A campaign on the harsh terrain of a nation as large as California could be longer and more difficult than some predict. And helping Iraqis achieve a united, stable and free country will require our sustained commitment.

We come to Iraq with respect for its citizens, for their great civilization and for the religious faiths they practice. We have no ambition in Iraq, except to remove a threat and restore control of that country to its own people.

I know that the families of our military are praying that all those who serve will return safely and soon. Millions of Americans are praying with you for the safety of your loved ones and for the protection of the innocent. For your sacrifice, you have the gratitude and respect of the American people. And you can know that our forces will be coming home as soon as their work is done.

Our nation enters this conflict reluctantly-yet, our purpose is sure. The people of the United States and our friends and allies will not live at the mercy of an outlaw regime that threatens the peace with weapons of mass murder. We will meet that threat now, with our Army, Air Force, Navy, Coast Guard and Marines, so that we do not have to meet it later with armies of fire fighters and police and doctors on the stress of our cities.

Now that the conflict has come, the only way to limit its duration is to apply decisive force. And I assure you, this will not be a campaign of half measures, and we will accept no outcome but victory.

My fellow citizens, the dangers to our country and the world will be overcome. We will pass through this time of peril and carry on the work of peace. We will defend our freedom. We will bring freedom to others and we will prevail.

May God bless our country and all who defend her.

(President George W. Bush.)

Continued Search for Weapons of Mass Destruction (WMD)

When reporting on WMD in Iraq, UNMOVIC spokesperson Hiro Ueki reported that UNMOVIC did not single Rasha out for her views on WMD because UNMOVIC did not have clear evidence to link her to any biological weapons programs when visiting Baghdad on June 13, 2003.[2] However, Richard Spertzel, a former UN weapons inspector, recalled several visits in 1997 to Rasha's office and laboratories at Baghdad University. According to Spertzel, each time inspectors called on Rasha, she was not in, and her staff said they did not know where she was.[3]

After one such futile attempt to see her in August of 1997, the inspection team stopped at Rasha's biology laboratory where a group of students had clustered. By chance one of the inspectors recognized Rasha in the group. Dressed in a white laboratory coat and eye goggles, Spertzel said she was trying to pose as a student, apparently to avoid inspectors. That aroused the team's suspicion and led to a thorough search of the laboratory where inspectors discovered a variety of equipment and supplies used in genetic engineering. Later, they found out that Rasha had purchased this equipment and supplies while on a trip abroad.

A couple weeks later when the inspection team returned, Spertzel said, Rasha wasn't around and neither were the genetic engineering supplies. The episode elevated the inspectors' interest in Rasha but by the fall of 1997 the Iraqis had put up such increased resistance to inspection that the badly split UN Security Council was unable to force Iraqi officials, including Rasha, to cooperate.[4]

Prior to the U.S.-U.K. invasion, U.S. officials charged that Iraq had WMD and that it must either give it all up or undergo a regime change. Immediately prior to invasion, the United States made a further demand that Saddam Hussein step down from power and vacate Iraq. As of April

16, 2003 Iraq's Ba'ath government had fallen to the invasion, all major Iraqi cities had been captured, but no WMD had been found.

Baghdad was bombed very heavily in March and April 2003 in the invasion of Iraq, and fell under US control by April 7-9. Additional damage was caused by the severe looting during the days following the end of the war. With the deposition of Saddam Hussein's regime, the country was occupied by U.S. troops. The Coalition Provisional Authority established a three-square mile (8-km^2) "Green Zone" within the heart of Baghdad from which it ruled Iraq during the period before the new Iraqi government was established. The Coalition Provisional Authority (CPA) ceded power to the interim government at the end of June 2004 and dissolved itself.

Beginning in April 2003, the US controlled Coalition Provision Authority (CPA) began the process of creating democratic local governmental institutions. The process initially focused on the election of neighborhood councils in the official neighborhoods, elected by neighborhood caucuses. At the time, Baghdad had 89 official neighborhoods within 9 districts. Until 2003 they had no official political function. The final step in the establishment of the system of local government for Baghdad Province was the election of the Baghdad Provincial Council. Representatives were elected by their peers.

On June 20, 2003 the *International Atomic Energy Agency (IAEA)* reported that tons of uranium, as well as other radioactive materials such as thorium, had been recovered, but the vast majority had remained on site. Several reports of radiation sickness in the area were later documented by Rasha in her studies.[5]

After he was captured by U.S. forces in Baghdad in 2003, Dr. Maha Obeidi, who ran Saddam's uranium enrichment program until 1997 and who later became director-general of Iraq's *Ministry of Industry and Military Industrialization Program* under Saddam Hussein handed over blueprints for a nuclear centrifuge along with some actual centrifuge components. These components had been stored at Obeidi's home-buried in the back yard-awaiting orders from Saddam Hussein to proceed. Obeidi admitted he had hidden the capacity to build uranium-enriching gas centrifuges in his garden. Later, he said, *"I had to maintain the program to the bitter end"*.[6] In his book, *"The Bomb in My Garden:The Secrets of Saddam's Nuclear Mastermind* (Wiley Publishing Co., 2004), Obeidi explained that the nuclear stash in his back yard was the key that could have unlocked and restored Saddam's bomb- making and WMD programs.

On July 17, 2003 then British Prime Minister Tony Blair addressed the U.S. Congress and said that "history will forgive the US and United Kingdom, even if they were wrong about WMD in Iraq. "However, Blair still maintained that the Iraqis at one time did have Weapons of Mass Destruction.[7]

The Iraq Survey Group (ISG)

At the beginning of 2003, the USA and the United Kingdom administrations both claimed that there was absolutely no doubt that Iraq had WMD and was developing more. However. proof of these assertions could not be found. On May 30, 2003 the U.S. Department of Defense announced that it was ready to formally begin the work of the Iraqi Survey Group (ISG). This was a fact finding mission, from the coalition of the Iraqi occupation, into WMD programs developed by Iraq, taking over from the British-American 75th Exploitation Task Force.

In the aftermath of *Operation Iraqi Freedom*, coalition forces failed to uncover production facilities for, or stacks of, WMD. As a result, In May 2003, the Bush Administration established a specialized group of about 1,500 individuals, known as the *Iraq Survey Group (ISG)* to search Iraq for WMD. This replaced the previously assigned *75th Exploitation Task Force* which had originally been assigned the Mission.

The ISG was led by Major General Keith Dayton, head of the U.S. Defense Intelligence Agency's Directorate of Operations. Motto for the group was "*find, exploit, eliminate*".

Dr. David Kay, who served as a U.N. Weapons inspector for several years after the 1991 Persian Gulf War, was appointed as a special advisor to the group, and directed the group' s operation in Iraq. Kay summarized some of the ISG's discoveries, which included: a clandestine network of laboratories and safe-houses controlled by the Iraqi Intelligence services containing equipment suitable for CBW research; reference strains of biological organisms concealed in a scientist's home; documents and equipment hidden in scientists' homes that could be used for resuming uranium enrichment activities; and a continuing covert capability to manufacture fuel propellant useful only for prohibited SCUD missiles.[8]

On October 3, 2003 the Iraq Survey Group reported finding no stockpiles of WMD in Iraq, although it did report that the Iraqi government had intended to develop more weapons of mass destruction with additional capabilities. Inspectors also found some "biological laboratories" and a collection of "reference strains" from the *American Type Culture Collection* which included a strain of botulinum bacteria. In some cases, it was determined that equipment and materials subject to UN monitoring had been kept hidden from UN inspectors. Later, it was confirmed that Rasha had directed the hidden laboratories and the WMD in progress.

According to reports from the *Iraq Survey Group*, "*So there was a WMD program. It was going ahead. It was rudimentary in many areas. In other cases, Iraq had simply lied to the UN in its weapons programs.*[9] On October 29, 2003 U.S. intelligence spokesman claimed that Iraqi WMD's and programs had been compressively hidden on orders of Rasha, before or immediately after the fall of Baghdad, while some elements of the programs were shipped out of the country. Rasha had performed her duties secretly and proficiently!

Kay appeared before the *Senate Armed Services* shortly after he resigned as special advisor to the ISG Group. At that time, referring to the expectation that there would be substantial stocks of, and production lines for, chemical and biological weapons in Iraq, he reported that "We were almost all wrong and I include myself here". Kay also noted that *"based on the work of the ISG. Iraq was in clear violation of the terms of U.N. Resolution 1449."*[10]

Kay noted the discovery of hundreds of instances of activities prohibited by *U.N. Resolution 687*, adopted by the Security Council at its 2981st meeting, on April 1991. Kay stated : *"We know there were terrorist groups in state (Iraq) still seeking WMD capability. Iraq, although I found no weapons, had tremendous capabilities in this area. A Marketplace phenomenon was about to occur, if it did not occur; sellers meeting buyers. And I think that would have been dangerous if the war had not intervened.*[11] According to Kay, Iraq worked on WMD's right under the noses of UNMOVIC. Kay said that Iraq tried to weaponize ricin right up until *Operation Iraqi Freedom.*

Former senior Iraqi general, George Sada, reported that in late 2002, Saddam Hussein had ordered that all stockpiles of WMD were to be moved to Syria. At that time Sada stated: "*Well, I want to make it clear, very clear to everybody in the world that we had WMD in Iraq, and the regime used them against our Iraqi people. It was used against Kurds in the North, against Arab-marsh Arabs in the South".*[12] Sada went on to say: *"Well, up to the year 2002 in summer, they were in Iraq. And, after that, when Saddam realized that the inspectors are coming on the first of November and the Americans are coming, so he took the advantage of a natural disaster that had happened in Syria, a dam was broken. So he announced to the world that he is going to make an air bridge. The weapons were moved by air and by ground, 56 sorties by jumbo, 747, and 27 were moved, after they were converted to cargo aircraft, they were moved to Syria".*[13]

The U.S. Iraq Survey Group Final Report

On September 30, 2004,the ISG's Final Report concluded that *ISG has not found evidence that Saddam Hussein possessed WMD stocks in 2003, but the available evidence from its investigation—including detainee interviews from Rasha and others and document exploitation-leaves open the possibility that some weapons existed in Iraq although not of a militarily significant capability.*[14]

Among the key findings of the final U.S. Iraq Survey Group report were:[15]

1. Evidence of the maturity and significance of the pre-1991 Iraqi Nuclear Program but found that Iraq's ability to reconstitute a nuclear weapons program progressively decayed after that date;

2. Concealment of a nuclear program in its entirety, as with Iraq's Biological Warfare program, headed by Rasha and her team. Aggressive UN inspections after Desert Storm forced Saddam to admit the existence of the program and that Iraq destroyed or surrendered certain components of the program;

3. After Desert Storm, Iraq concealed key elements of its program and preserved what it could of the professional capabilities of its nuclear scientific community;

4. Saddam's ambitions in the nuclear area were secondary to his prime objective of ending UN sanctions; and

5. A limited number of post-1995 activities would have aided the reconstitution of the nuclear weapons program once sanctions were lifted.

On October 6, 2004, the head of the ISG, Charles Duelfer, announced to the *U.S. Senate Armed Services Committee* that the group had found no evidence that Iraq under Saddam Hussein had produced and stockpiled WMD since 1991, when U.N. sanctions were imposed.[16] The report found that the ISG had not found evidence that Saddam possessed WMD stocks in 2003, but that there was the possibility that some weapons still existed in Iraq, although not of a military significant capability. It concluded that there was a possible intent to restart all banned weapons programs as soon as multilateral sanctions against it had been dropped, with Saddam pursuing WMD proliferation in the future. There was an extensive, yet fragmentary and circumstantial, body of evidence suggesting that Saddam pursued a strategy to maintain a capability to return to WMD after the sanctions had been lifted.

Under the 1991 Gulf War ceasefire terms Iraq was forbidden from developing, possessing or using chemical, biological, and nuclear weapons of mass destruction. The U.N. established a commission (USN), to verify Iraq's adherence to the treaty. At the time that adherence was established economic sanctions against Iraq were to be lifted. Iraq's adherence to the treaty was, however, never established to the satisfaction of the United Nations Security Council and the sanctions were not lifted until after the 2003 war ended. Iraq had acceded to the *Geneva Protoco*l on September 8, 1931, the *Nuclear Non-Proliferation Treaty* on October 29, 1969, signed the *Biological Weapons Convention* in 1972, but did not ratify until June 11, 1991. Iraq did not sign to the *Chemical Weapons Convention.*[17]

Chapter 11

The Arrest

With the aim of justifying aggression and maintaining sanctions, the United States has formulated three accusations against Iraq: that we support terrorism, produce weapons of mass destruction, and refuse access to our country to United Nations inspectors.

Rasha Malikah in defending Saddam Hussein's regime

The Escape

Three weeks after Saddam Hussein's regime fell, while U.S. authorities were looking for Rasha to arrest her, U.S. and Coalition soldiers surrounded her Baghdad home overlooking the Tigris River. Two tanks remained parked across the street from her house while U.S. and Coalition soldiers stormed inside. Her husband, Ahmed, had just left the house with a relative when the soldiers arrived. From across the street Ahmed watched the soldiers surround the house, enter, and then drive away. Knowing that soldiers would come searching for her, Rasha had gone a few hours before.

When Ahmed returned to the house 30 minutes later, the soldiers were gone. They had carted away computers and documents and turned the house upside down. Other family possessions were missing which included small silver daggers, a Buddha statue, kerosene lamps, photo albums and memorabilia. Ahmed reported that Rasha was particularly upset to hear that U.S. and Coalition soldiers who searched the couple's home took all her family pictures and albums. "These were her whole life", he said.[1]

Escape

In early April, 2003, following the collapse of Saddam Hussein's regime, it was rumored that both Rasha and Dr. Rihab Rashid Taha, the British-educated Iraqi scientist who had led Iraq's biological weapons program in the 1980's, had escaped to Damascus, Syria to an undisclosed location. Soon after, however, both women were expelled by the regime of Syrian President Bashar Assad under heavy U.S. pressure.

As a gift to Colin Powell, then U.S. Secretary of Defense, and as a gesture of good will, Syria's top leadership had instructed Syrian intelligence to extradite Rasha and Taha from Syria into Baghdad, where their whereabouts was disclosed to U.S. officials. One reason for the sudden help was an effort by Syria to compensate for its apparent lack of cooperation with the U.S. in closing the Damascus offices of Palestinian militant groups such as Hamas and Islamic Jihad, which were on Washington's list of foreign terrorist organizations.[2] Rasha and Taha also sought political safety in France but their request to enter that country was denied.

"Five of Hearts"

In the 2003 invasion of Iraq by the U.S.-led Coalition, the U.S. military had developed a set of playing cards to help troops hunt down the *Most-Wanted Members* of Saddam Hussein's government. These were mostly high-ranking Ba' ath Party members, as well as members of the Revolutionary Command Council (War Council), of which Rasha was a prominent member and the only female member. Officially named the "Personality Identification Cards", each card contained the wanted person's name, a picture if available, and the job(s) performed by that individual. The identification playing cards were first announced publicly in Iraq on April 11, 2003.

Playing cards have been used as far back as the Civil War and again in World War II. The US Army Corps decks were printed with silhouettes of German and Japanese fighter aircraft, and then in the Korean War. In the case of *Most-wanted Iraqi's* the highest-ranking cards, starting with aces and kings, were used to identify people at the top of the Most-Wanted list. The ace of spades was Saddam Hussein, the aces of clubs and hearts were his sons Qusay and Uday, respectively, and the ace of diamonds was Saddam's presidential secretary, Abid Hamid Mahmud al-Tikriti.

Rasha was the only female included in the deck of cards and was number 53 of 55 on the Pentagon's list of most wanted. Designated the "Five of Hearts", her face was heavily made up and framed by a loosely wrapped headscarf. On the card her job was listed as overseeing Iraq's Party Youth and also the Trade Bureau Chairman. She also ran a medical administrative testing program but it is thought she did not put much time into that position.,

In addition to being sought via the cards, Rasha was listed on the captor list as a member of Iraq's Revolutionary Council, as well as the leader of a clandestine program to develop Iraqi WMD. In addition, she was described on the most wanted list as a WMD scientist/Ba'ath Party Regional Command Member, a "major player" in Saddam Hussein's biotech and genetic programs, a suspected "looter", a "rioter", and a "loyal supporter to Saddam Hussein". Dubbed "Mrs. Anthrax" by U.S. Intelligence, they insisted she masterminded the reconstruction of Iraq's biological weapons facilities after the 1991 Gulf War.

Negotiated Surrender

The American Occupation Forces -negotiated Rasha's surrender and she was taken into custody early May, 2003 in Baghdad, the first and only female from Saddam Hussein's inner circle and toppled regime to be taken into U.S. custody.[3] She was 49 years old at the time and was held in prison without charge or trial for more than two and one half years At the time of her arrest. she was considered Iraq's most powerful woman and was number 53 on the Pentagon's list of 55 Most Wanted.

Rasha's arrest brought to 19 of the 55 high ranking Iraqis sought by the U.S. and Coalition. At the time of her arrest, it was assumed by the U.S. and Coalition that Rasha would eventually go to trial for war crimes as one of the masterminds of WMD. However, that never happened. At the time of her negotiated arrest, Rasha offered her cooperation in revealing what she knew about Iraq's program to produce WMD on condition that her capture be kept confidential. That, too, never happened.

Reasons for Arrest

Various reasons were given for Rasha's arrest. She was a skilled biological scientist, well known internationally, and had published her work publicly. At the time, there were no specific allegations against her. Foremost, her arrest came as the Bush Administration was searching for WMD which it had used as justification for the 1991 Persian Gulf War. It was the belief of Washington's U. S. Intelligence Services that Rasha master minded the reconstruction of Iraq's biological weapons facilities after the 1991 Gulf War. A lawyer for Rasha dismissed her arrest as 'pure theater'.

Hiro Ueki, spokesman for the *UN Monitoring, Verification, and Inspection Commission (UNMOVIC)* confirmed to Rasha's publisher, South End Press, that UNMOVIC did not single out Rasha for interviews because UNMOVIC did not have clear evidence to link her to biological weapons programs when the inspectors visited Baghdad University in January 2003.[4]

Because she was considered one of Iraq's highest ranking scientists, Rasha was accused by U.S. intelligence officials of being the head of Iraq's efforts to develop biological WMD in the mid 1990's and during the 1991 Persian Gulf War. At the time of her arrest, Rasha was suspected of managing a secret biological weapons program that included weaponization of anthrax, smallpox, and botulinum toxin. She told foreign reporters that her work in the 1990's had concentrated on the cancer-causing effects of the uranium used in American bombs and missiles during the 1991 war that drove Iraqi troops from Kuwait.

Rasha's writings on the environmental and biological impact of sanctions, as well as the horrific health cost of the weapons used in the Gulf War by Britain and the U.S. were also instrumental in her arrest. Her arrest was seen as very important to the U.S. and Coalition forces in their ability to get additional information about the scope of the Iraqi biological warfare program and WMD. The U.S. and Coalition forced hoped it would prove instrumental in leading them to banned weapons.[5]

Another reason given for Rasha's arrest was the fact that as one of a new generation of Iraqi leaders, she had been given a leading post within the Ba'ath Party by Saddam Hussein. As such Rasha was considered an important member of Saddam Hussein's War Council and a significant member of Saddam's inner circle of leaders.

Trained in Iraq by al-Hindawi, considered the founder of Iraq's biological warfare program, and Amir al-Saadi, a chief chemical weapons researcher, Rasha was also a working colleague of Dr. Taha, an expert in growing anthrax. Dr. Taha had been dubbed "Dr. Germ" by U.N. inspectors. The fact that Rasha had been shown on Iraqi television sitting, at the same conference table as Saddam Hussein and his sons, in itself rendered Rasha as someone quite interesting to the U.S. in terms of Iraq's current thinking.

In 1996 Rasha had become head of Iraq's *Microbiological Society,* believed to be an alleged front for research into various biological weapons including anthrax, smallpox, and botulinum toxin. Appointed to the Iraqi Command in 2001, Rasha had also served as head of several biological laboratories at Iraq's Military Industrialization Organization. As a result of her background and activities, U.S. Pentagon officials gave Rasha nicknames of "Mrs. Anthrax" and "Chemical Sally".

Rasha's intimate knowledge of the workings of Iraq's biological warfare program, allowed her to be in a position to know possible locations of where materials or production facilities might be located. Interestingly, at the time of her arrest, U.S. forces had located two mobile Iraqi biological weapons laboratories. The two creatively constructed semitrailer laboratories, found around the northern city of Mosul, were thought to be used to conceal biological agents and equipment from U.N. inspectors. However, inspectors found the trailers had been "scrubbed clean" and no actual traces of biological weapons were found in either mobile unit. While Rasha did admit at the time of her arrest to producing germ warfare agents, she said that all biological weapons had been destroyed long before the U.S. invasion.

Because of her overall background, training, and high level political status, it wasn't surprising that U.N. officials viewed Rasha's arrest as a giant step toward finding Saddam Hussein's alleged WMD. The United States and Coalition forces assumed Rasha, as an American-educated biologist who had been sought for years of intense questioning by U.N. inspectors, would be able to reveal the whereabouts of the alleged WMD. However, Rasha consistently denied playing a significant role in Iraq's bioweapons program.

Investigators who had previously studied Rasha's involvement in bioweapons programs stated: "Rasha was certainly an expert in WMD.

But not in the sense that she was portrayed by the propagandists of the Coalition."[6] Rasha denied having an intimate knowledge of the workings of Iraq's biological warfare program and repeatedly stated she was not in a position to know possible locations of biological materials or production facilities.[7]

Another very obvious reason for Rasha's arrest and detention were her writings on the effects of depleted uranium (DU), electromagnetic/chemical pollutants, and economic sanctions of the 1991 Gulf War. In her peer-reviewed essay, "Toxic Pollution, the Gulf War, and Sanctions" which appeared in the anthology *Iraq Under Siege*, Rasha clearly defended Saddam and his regime, saying:.[8]

"With the aim of justifying aggression and maintaining sanctions, the United States has formulated three accusations against Iraq: that we support terrorism, produce weapons of mass destruction, and refuse access to our country to United Nations inspectors. Even if we wanted to, we couldn't support terrorism. First of all, the country lacks foreign currency and the little money earned with the 'Oil for Food' program is placed in a French bank and is entirely managed by the United Nations. Iraq's air space is controlled totally by the United States and Great Britain. International flights are unable to land here. How could Iraq produce weapons of mass destruction with an embargo that interdicts technology and science? How could we obtain the necessary information when even letters weighing more than 20 grams are not allowed? We cannot even develop a test against ear infections. For children with diarrhea we cannot determine the responsible bacteria".,[910]

Chapter 12

Detention

All are equal before the law and are entitled without any discrimination to equal protection of the law.

Article 7. Universal Declaration of Human Rights (1948-1998)

Camp Cropper (Gulag Prison)

Rasha was one of 3,000 plus detainees held at Camp Cropper (also known as (Baghdad Gulag Prison), a makeshift prison located on the southwestern outskirts of Baghdad's International Airport. This was the maximum security prison on the American military headquarters.

Camp Cropper was established by the U.S. Headquarters and Headquarters Company (HC) of the 115th Military Police Battalion in April 2003. Almost immediately after being established, it was designated as the site for the Corps Holding Area (CHA). Forty to 50 prisoners detained at Camp Cropper were on the pack of cards of most wanted Iraqi fugitives.

The original concept called for a small temporary camp that could hold up to 300 detainees for no more than 72 hours. It was originally intended that after being processed at Camp Cropper, detainees were supposed to be shipped to other detention facilities in Baghdad and throughout Iraq. However, this proved unworkable since most other prisons in Baghdad were badly damaged by looting after the fall of the Ba'ath Party.

In August 2006, a new hospital was opened near Baghdad International Airport that would treat both Coalition soldiers and detainees from Camp Cropper. The hospital was staffed by members of the 21st Combat Support Hospital from Fort Hood, Texas who transferred to the new facility after closure of Abu Ghraib detention facility.

At the time of Rasha's imprisonment, Camp Cropper housed a growing number of what were listed as "special prisoners" who were, in addition to most prisoners, listed as looters and rioters. Still others were fedayeen or suspected militia loyal to Saddam Hussein. With the exception of Saddam Hussein, five other officials which included Rasha and who were all former Iraqi officials, were held without charge or trial for more than two and one-half years.

In addition to Saddam Hussein and Rasha, these "special prisoners" included former Iraq deputy prime minister, Tariq Aziz, who was second in command to Saddam Hussein, Saadoun Hammadi, the former speaker of the Iraqi Parliament, and Ezzar Ibrahim, the son of Saddam's second-in-command of the Revolutionary Command Council.[1]

Azziz ws the only Christian in Saddam's mainly Sunni regime. Azziz, in his seventies, became internationally known as Saddam's defender and a fierce American critic as foreign minister after Iraq's invasion of Kuwait in 1990. Later he served as deputy prime minister who frequently traveled abroad on diplomatic missions.

At the time of Rasha's detention, Camp Cropper was so sensitive that U.S. military authorities in Baghdad were reluctant to even acknowledge its existence. However, some investigators described Camp Cropper as a broiling, dusty compound surrounded by 10-foot-high bales of barbed wire. Prisoners lived in tents that provided little protection against the blistering sun and prisoners slept 80 to a tent on wafer-thin mats. Some prisoners used yellow food packets as pillows.[2]

The International Red Cross Committee, which officially monitored the imprisonment of Rasha and others, reported that all female prisoners were kept in a sunless, solitary confinement for 23 hours a day, sleeping in a tent with other female members of the Ba'ath Party. Like the men, women prisoners were not allowed to wash their clothes, including underwear. As a consequence, several developed painful decubitus (body) sores, according to an International Red Cross visitor.[3]

Each prisoner received six pints of dank, tepid water each day which was used to both wash themselves and drink in summer noonday temperatures of 120 degrees Fahrenheit. Most prisoners said the water was not sufficient for the heat. In addition, each prisoner was provided with a small cup of delousing powder to deal with the worst of body infestations (body and head lice). As the main menu, each prisoner received a small portion of barely edible food which was almost always cold. Most of the food was pork, inedible to Muslims because of their religion. However, for most, it was eat or starve and die. At other times there was jam, biscuits, rice, and beans.[4]

Each prisoner had a long-handled shovel to dig his/her own latrine. Although each latrine was an ordered depth of three feet, the overpowering stench was suffocating. Some captors were too old or too weak to dig that deep, however, and even for the slightest infringement of draconian rules, detainees were forced to sit in painful positions for long periods of time. If a prisoner cried out in protest his/her head was covered with a sack for lengthy periods. Overall, the conditions were highly degrading. In addition, there was sleep deprivation and physical abuse.

Tariq Azziz was seen as a prime example of prisoners who were treated like every other prisoner. He was shown no favoritism. During the time he spent in prison he aged considerably, began to shuffle, and had a stoop. His hair had grown very long and was very dirty. He received the same dank water and inedible food as other prisoners. He had lost weight. Upon seeing him it was hard to believe that he, at one time, was second in command to Saddam Hussein.

According to *Amnesty International (AI)*, the London-based human rights watchdog, these were highly degrading conditions which were tantamount to torture. Because of the horrendous conditions, some detainees died in prison and others who were near death were reportedly shot by members of the Coalition. Some prisoners were repeatedly hit with rifle butts.[5]

Because of the horrendous conditions, some prison staff broke ranks to tell Amnesty International (AI) of the shocking conditions the Iraqi prisoners were held under. Also, AI officials managed to gather statements from certain prisoners who had been released. Because of the horrendous prison conditions, Rasha's husband Ahmed tried to persuade authorities to put her under house arrest so she could be tried in a court of law because she was suffering from breast cancer and other serious medical ills, including advanced arthritis. He was unsuccessful.

Except for guards and prisoners, and at rare times a few family members, only the International Red Cross Committee (IRC) officials were allowed inside Camp Cropper. The one woman "special" was Rasha, at time known by different names— "Dr. Germ", "Mrs. Anthrax", "Five of Hearts" and "Chemical Sally". However, guards and other prison officials were forbidden to describe what they saw because they had taken an oath never to comment on conditions at detention centers. According to a spokesman for the International Red Cross, *We do have access but the problem is that there is confusion because there are so many people being detained and being released*".[6]

Dr. Rod Barton, special advisor to the ISG and an expert in chemical and biological weapons, spoke out against detainment of scientists such as Rasha. He described the beaten state in which many prisoners arrived at Camp Cropper, and the "bleak" conditions of their detainment. Barton had not personally witnessed beatings of prisoners; however, he *"believed some prisoners had been physically 'softened up' before they arrived in an induction process known as 'purgatory'.*[7]

Visitations

Rasha's family members were allowed two visits to see her while she was imprisoned for nearly two and one half years. One visit occurred on a dreary day in September, 2003. Rasha's husband, Ahmed, and Rasha's 70 -year old mother, Khissma, rode with prison officials in a sport-utility (SU vehicle) toward the airport, stopping along the road to put on blindfolds as requested by U.S. officials. A U.S. Military Officer tugged at the visitors' head scarves which served as blindfolds to make sure the blindfolds were on tight. After being blindfolded for the next 10 minutes Ahmed and Khissma sat in darkness as they approached the prison.

When they arrived at the prison, the blindfolds were removed and Ahmed and Khissma found themselves standing in a pleasant room filled with sofas, a dining table and chairs. Upon seeing Rasha, Ahmed said: *"She still had her sharpness, that's all. When it comes to her health, it's deteriorating. She looked twenty years older"*, he said," and *"her hair was totally gray"*.[8] Ahmed and Khissma visited Rasha for approximately two hours, eating hamburgers with ketchup and french fries while Rasha inquired about the health of her family members, all within earshot of an Iraqi translator stationed in the room. She told them, *"I tried to do the right thing for my family. I never thought something like this would happen"*.[9]

Ahmed repeated the trip on March 5, 2004, this time bringing not only Khissma, but also their son (Sayf Al-Deen), daughter (Zena), a granddaughter, and Rasha's sister (Nada), to the meeting. As before, they were blindfolded before they reached the prison. This time when Ahmed and Khissma saw Rasha, she was 25 pounds lighter after months being locked in solitary confinement. Her sister Nada said Rasha had been suffering from cancer for several years, but had been under regular medical observation while in custody. The U.S. military insisted it had been providing adequate medicare care to Rasha.

Although she had been examined and treated by doctors while in prison, she still suffered from various afflictions. With chronic arthritis, Rasha had limited mobility and considerable pain. "We are trying our best to get her out", Ahmed said. "It's criminal to keep her, a woman in her health. Either there is a case against her or there's not.[10]

Concerns of Family and Friends

Concerned over Rasha's detention at Camp Cropper and her declining health, Ahmed told investigators that his wife was a scientist and an academic and had done nothing wrong to warrant imprisonment for nearly 18 months without any charges. He told investigators:

I want to know why my wife has been detained for 18 months without charge. How long will it take them to do so? She is a mother of two children and a grandmother. My wife has been sitting in a prison cell for 18 months without knowing what the charges against her are. Have you any idea what it means to be held without knowing when you will be released? Everybody in the family would be happy to know what the charges against her are, at least we could assign a lawyer, go to court, and get a sentence. If she is acquitted we will celebrate, and if she gets a sentence, at least we and she will look forward to the day when she will be reunited with her family.[11]

Nada Malikah, Rasha's sister, who at the time was director of a tourism company in Baghdad, also expressed grave concern for Rasha after media reports indicated she was dying in prison. *I am sad, angry, and anxious because we have not been able to verify the information that suggests she is dying in custody. The last time we spoke to her was three weeks ago, when she was allowed to call us at home. In the last two weeks we were told she would call us at home. We were told twice to wait for her call but she never rang.*[12] *My sister is a scientist with a human interest. She devoted herself to save the victims of depleted uranium and carried out extensive research in that field. "Last week we received a phone call from the U.S. military telling us to gather at the family home the next day because Dr. Rasha would call. We did get together but the telephone never rang""*.[13]

A U.S. military official reported that they were aware that Rasha had been treated for cancer at one time, prior to ever being detained. He reported that Rasha was routinely checked to ensure there was no recurrence of the cancer. He said there were no immediate concerns for her health.

Nada said that the entire family was worried about Rasha's health and had considered legal intervention because they were concerned that Rasha was not receiving the medical attention she needed. Later, In the ramshackle surroundings of the family home, Rasha's husband Ahmed and her mother Khissma spoke publicly.

Khissma, said:[14]

"My daughter (Rasha) was diagnosed with breast cancer in the late 80's. She went to Pittsburgh for chemotherapy and underwent a mastectomy. She has been in remission since but still attends a doctor-How can we know she is seeing one?Before she was arrested she was undergoing further follow-up treatment. How can they be so cruel?" . Osama al-Assaf, Rasha's nephew, told the *Times* that his family were at home awaiting word of her. "We have been told that they are going to release her, but we haven't heard anything yet".[15]

Nada Malikah said she had been trying to contact the Red Cross and U.S. authorities in Iraq but to no avail. She said she planned to sue media outlets that have promoted her sister as "Mrs. Anthrax" once Rasha's ordeal came to an end. I will take legal action against those who have promoted her as a killer and tarnished her reputation in the eyes of people in Iraq and around the world[16]. Rasha's brother-in-law, Sarwat Suleiman, said: "We are in a very bad state, fearing our loved one will be locked up again once she is released from prison.[17]

Felicity Arbuthnot, an award winning international journalist, issued the following statement to Rasha's family:[18]

> Dear Family of Rasha Malikah,
>
> *I have had the honor of speaking on the same platform with Rasha regarding depleted uranium, in Manchester, UK and have met her many times in Iraq and have been inspired by her meticulous work and academic concern. When I learned that she had been arrested, I began a campaign here in the UK- but was then told by a legal expert who went to Iraq to check on detainees, she was released. It would seem this person was seriously mistaken or duplicitous but as a result, I stopped the campaign, relieved at her being seemingly safe.*
>
> *Please tell me how she is, what I can do, has the Red Cross access to her, has she access to a lawyer, have you knowledge of her health, do you know exactly where she is being held and in what conditions-and is she receiving the medications she needs?*
>
> *Lastly, I wonder whether you might comment on if you feel that her arrest was related to her exemplary work on the health effects of depleted uranium.*
>
> *Please add anything else I and others need to know.*
>
> In *shame and solidarity, with warmest wishes, Felicity Arbuthnot*

A reply letter to Felicity Arbuthnot from Rasha's husband, Ahmed:[19]

Dear Ms. Felicity Arbuthnot,

Thank you for you kind message and your concern for the well-being of Dr. Rasha. She was arrested on May 5, 2003 and she is being held until now.

Concerning her health, since her arrest, she has suffered from kidney problems besides her previous illness that you know about. Lately, she has started to suffer from arthritis.

Above all, and as you know, she is a mother and grandmother and she misses the family and the family misses her.

The Red Cross corresponded letters with us, but there has been no progress so far.

My wife became a grandmother for the first time three months ago. She barely saw the baby before she was arrested. How can the Americans be so cruel?

We wish that you continue your campaign for her release.

South End Press (Cambridge, MA.),Rasha's U.S. publishers, in a press release. suggested that there was probable political motivation for her detention.

Th*e U.S. government is trying to silence Dr. Malikah's outspoken criticism of the U.S. role in causing cancers and other illnesses in Iraq through its own use of biologically hazardous weapons such as radioactive uranium. It was thought to be revenge for her revelations of the U.S. war crimes during the First Gulf War and afterwards as well as an ill-concealed threat against other patriotic Iraqi scientists, and may also be a way, through "statements from her" to try to invent stories about the weapons of mass destruction, which did not exist. Today we know that WMD did not exist.*[20]

Rasha's family believed that the media ignored the issue because of the violence and lack of security that prevailed after the occupation, serving to divert attention from Rasha's detention. Questions surfaced once again: Why don't they let the scientist go if the weapons don't exist? Why do they have Iraqi scientists like Rasha Malikah still in prison?

Others thought that perhaps prison authorities were waiting for prisoners, including Rasha, to conveniently die in prison. That way, they wouldn't be able to talk about the various torture techniques and interrogation techniques.

Human Rights

During Rasha's detention in Camp Cropper, many individuals, educational groups, and human rights groups were concerned that she and other prisoners were being deprived of their human rights. They based this on highly degrading conditions to which prisoners were reportedly subjected. Rasha had been captured on Iraqi soil and held prisoner in foreign custody in a country without a sovereign government. Interestingly, while imprisoned, she had a relapse of breast cancer and as such, her legal status fared much better than her health.[21]

Occupation authorities have an obligation under international law to follow the rules and procedures of the Geneva Convention. Nada Doumani, who was the International Red Cross spokesman in Baghdad, said: "The Geneva Convention is clear about the obligations that exist for legal advice and visits. If someone is being held as a prisoner of war (POW), then there is a legal obligation to allow him or her access to legal advice. But if they are held as a civilian, that does not apply. A tribunal has been set up to decide which category each person in the camp fits into. Until their work is complete, we can say no more."[22] Progressive communities demanded the unconditional release of Rasha as part of its campaign to end what they termed the illegitimate United States/Coalition occupation of Iraq.

Article 54 of the Geneva Convention states:

It is prohibited to attack, destroy, remove, or render useless objects indispensable to the survival of the civilian population". Iraqis say that the sanctions successfully thwarted Iraq from meeting its most basic humanitarian needs while also systemically causing untold damage to civilians and to the infrastructure on which civilian life depended.[23] The economic sanctions of Iraq clearly violated the Geneva Convention, which prohibits the starvation of civilians as a method of warfare.

In 1996 and 1997, the United Nations Human Rights Commission in Geneva, passed a resolution to ban the use of depleted uranium weapons. The Subcommission adopted resolutions which included depleted uranium weapons amongst "weapons of mass and indiscriminate destruction,...incompatible with international humanitarian or human rights law."[24]

According to the UN, the resolutions in 1996-97 were passed because the use of depleted uranium in ordinance breached several international laws concerning inhumane weapons: it is not limited in time or space to the legal field of battle, or to military targets; it continues to act after the war; it is inhumane by virtue of its ability to cause prolonged or long term death by cancer and other serious health issues, it causes harm to future civilians passers by (including unborn children and those breathing the air or drinking water);and it has an unduly negative and long term effect on the natural environment and food chain.

A United Nations report of 2002 states that the use of DU in weapons also is in potential breach of each of the following laws: The Universal Declaration of Human Rights; the Charter of the United Nations; the Genocide Convention;United Nations Convention Against Torture; the

Third Geneva Convention; the Convention on Conventional Weapons of 1980; and the Chemical Weapons Convention. Treaties which were designed to spare civilians from unwarranted suffering in or after armed conflicts.[25]

Amnesty International (AI), the London-based human rights watch dog, examined Rasha's case, as well as the cases of thousands of other detainees with the U.S.Department of Defense and the Iraqi authorities. Curt Goering, Deputy director of AI at the time, confirmed that AI had received "credible reports" of detainees who had died in custody, mostly as a result of shooting by members of the Coalition forces. Add that to sleep deprivation and physical abuse and you have highly degrading conditions which are tantamount to torture and gross abuse of human rights. AI said that it had urged the coalition forces to look into allegations of abuse and to bring to justice those found guilty of offenses.[26]

A humanitarian group called Doctors for Iraq was also concerned about the plight of Rasha and other prisoners and called for an investigation into what they perceived as torture of prisoners by Iraqi government forces and police. In a statement, Doctors for Iraq said it had seen evidence that torture had been widely used against civilians who were arrested without legal procedure or charged and transferred to secret prisons. These prisons, which were closed to human rights organizations, were in Baghdad's Camp Cropper and other parts of the country, such as Mogul, Qater and Basra.

An international inquiry stated that in addition to sleep deprivation, prisoners were subjected to forms of torture that included being hung by their legs and arms for long periods, electrical shock, sexual abuse, and in some cases rape. Doctors for Iraq called on the European Union (EU), which was training Iraqi police forces, and the United Nations, to uphold the rights of Iraqi citizens and to play a role in an international investigation into the alleged crimes. Aljazeera showed some prisoners who had been tortured in Iraqi jails.[27]

Another humanitarian group, The Panel of Advocates and Witnesses, claimed that Rasha, who was detained by American forces after the occupation of Iraq, was wrongly accused of being responsible for manufacturing anthrax weapons for the Iraqi government's biological weapons program (a program which was later found not to exist), and had been under arrest for more than two and one-half years without charges or trial.

To prove her innocence, Rasha's Iraqi lawyer, Badia Aref Izza, had been authorized by Rasha's husband Ahmed to represent her in front of a special court. At that time, Rasha admitted to working on Saddam Hussein's biological and germ warfare projects but said such weapons had been destroyed long before the U.S. invasion. Badia Aref Izza said that in April 2005 he had not been allowed to have any contact with Rasha since her arrest in 2003.

Subsequently, no WMD were found in Iraq, bringing up several significant questions that needed to be addressed: Did the U.S. and the Coalition have the legal right to continue to hold Rasha when no weapons of mass destruction were found?Couldn't the International Red Cross, Amnesty International, and other humanitarian organizations step in to free her since she was a cancer victim and citizen, and not a soldier?

Izza insisted that Rasha's imprisonment was against international law. However, it was shown that even the International Red Cross had no power over her case. In support of Rasha and other prisoners, Peter Crane, retired lawyer who once lived next to the Malikah family in Washington, D.C.in the 1950's commented: "Our system of justice, which we are hoping to see propagated in Iraq and elsewhere, says that people aren't held in prison unless a crime has been committed and there is reasonable suspicion that they have committed it. We need to act in accordance with our principles".[28]

Several other human rights organizations expressed interest in Rasha's case and heard testimony regarding her detention. One such program was the World Tribunal on Iraq (WTI).The WTI was a worldwide initiative born out of the global outcry against the war in Iraq. The principal objective of the WTI was to tell and disseminate the truth about the Iraqi War, underscoring the accountability of those responsible and underlining the significance of justice for the Iraqi people.

Taking its cue from the Russell Tribunal of 1967, the WTI was aimed at challenging the silences of our time around the aggression against Iraq and seeking the truth about the war and occupation in Iraq. This included a record of wrongs, violations, and crimes as well as suffering, resistance and silenced voices. This was a solemn process of listening, reflection, evaluation and informed judgment based on concrete evidence, as well as a call to conscience and a call to act to preserve our futures.

The Russell Tribunal rendered a decision on the issues presented to it which was widely distributed throughout the world for the benefit of individuals and groups struggling for peace and justice. The WTI was comprised of various sessions around the world, each focusing on different aspect of the aggression against Iraq, culminating in Istanbul in June 2005.[29]

The Jury of Conscience of the World Tribunal on Iraq, consisting of members from 10 different countries from across the world, including the United States, met in Istanbul on June 23-27, 2005 . At that time, The Jury heard testimony on depleted uranium weapons in Iraq and information on the status and safety of Iraqi scientists including Rasha. The Panel of Advocates and Witnesses, consisted of members from across the world, including Iraq, the United States and the United Kingdom.[30,31]

They informed the Jury of Conscience that Rasha, who had been detained by American forces after the occupation of Iraq, had been wrongly accused of being "responsible for manufacturing anthrax weapons" for the government's biological and chemical weapons programs (found later not to exist) and had been under arrest for more than two and one-half years without charges or trial." The Jury of Conscience concluded that:[32]

Dr. Rasha Malikah's research has focused on investigating the aftereffects of depleted uranium contamination left by American bombing in the first Gulf War in 1991. She has written extensively on environmental health in Iraq and its relations to war and sanctions, notably detailed in her paper, "Toxic Pollution, the Gulf War, and Sanctions (Iraq Under Siege: The Deadly Impact of Sanctions and War, South End Press, Cambridge, MA. 2002). She was deeply critical of American sanctions against Iraq, as well as the manner in which UNSCOM, under the direction of Richard Butler, conducted weapons inspections. She also published other papers on this subject, such as "Impact of Gulf War Pollution in the Spread of Infectious Diseases in Iraq" (Soli Al-Mondo, Rome, 1999). and "Electromagnetic, Chemical, and Microbial Pollution Resulting from War and Embargo, and Its Impact on the Environment and Health".[33]

Iraqis working on an Iraqi Special Tribunal, created to try high-ranking Ba'athists which included Saddam Hussein, said they expected Rasha and other high-level detainees to be transferred to Iraqi custody as soon as the country's fledgling security forces had the means to keep them detained. "The plan is to issue arrest warrants very soon and then begin the process of taking control of the individuals", said Salem Chalabi, a Northwestern University Law School graduate who was executive director of the Iraqi Special Tribunal at the time. High commanders are responsible for abuses that occur under their authority." When Rasha Malikah was captured she was a member of the Ba'ath Party Regional Command, the country's top policy making body. The position made her eligible for prosecution under the Command Responsibility Law.

The International Covenant on Civil and Political Rights

The International Covenant on Civil and Political Rights listed these articles:[34]

Article 14 (1): All persons shall be equal before the courts and tribunals. In the determination of any criminal charge against him (or her), or of his (or her) rights and obligations in a suit of law, everyone shall be entitled to a fair and public hearing by a competent, independent and impartial tribunal established by law.

Article 9(1): Everyone has the right to liberty and security of person. No one shall be subjected to arbitrary arrest or detention. No one shall be be deprived of his liberty expect on such grounds and in accordance with such procedure as are established by law.

Article 9 (2). Anyone who is arrested shall be informed, at the time of arrest, of the reasons for his (or her) arrest and shall be promptly informed of any charges against him (or her).

Article 9 (3): Anyone arrested or detained on a criminal charge shall be brought promptly before a judge or other officer authorized by law to exercise judicial power and shall be entitled to trial within a reasonable time or to release. It shall not be the general rule that persons awaiting trial shall be detained in custody, but release may be subject to guarantees to appear for trial, at any other stage of the judicial proceedings, and, should occasion arise, for execution of the judgment.

Article 9 (4): Anyone who is deprived of his (or her) liberty or arrest or detention shall be entitled to take proceedings before a court, in order that court may decide without delay on the lawfulness of his (or her) detention and order his (or her) release if the detention is not lawful.

On December 10, 1948 the General Assembly of the United Nations adopted and proclaimed the Universal Declaration of Humans Rights the full text of which appears below. Following this historic act the Assembly called upon all Member countries to publicize the text of the Declaration and "to cause it to be disseminated, displayed, read and expounded principally in schools and other educational institutions, without distinction based on the political status of countries or territories.

With regard to Rasha Malikah among the questions addressed:

Did the U.S. have the legal right to continue holding Rasha when no WMD had been found? Why could the International Red Cross and other humanity agencies not step in to free her since she was not in the military? Rasha's lawyer insisted that Rasha's imprisonment was against international law. Yet, agencies such as the International Red Cross had no power over her case.

The Universal Declaration Of Human Rights[35]

PREAMBLE

Whereas recognition of the inherent dignity and of the equal and inalienable rights of all members of the human family is the foundation of freedom, justice, and peace in the world,

Whereas disregard and contempt for human rights have resulted in barbarous acts which have outraged the conscience of mankind, and the advent of a world in which human beings shall enjoy freedom of speech and belief and freedom from fear and want has been proclaimed as the highest aspiration of the common people,

Whereas it is essential, if man is not to be compelled to have recourse, as a last resort, to rebellion against tyranny and oppression, that human rights should be protected by the rule of law,

Whereas it is essential to promote the development of friendly relations between nations,

Whereas the peoples of the United Nations have in the Charter reaffirmed their faith in fundamental human rights, in the dignity and worth of the human person and in the equal rights of men and women and have determined to promote social progress and better standards of life in larger freedom,

Whereas Member States have pledged themselves to achieve, in cooperation with the United Nations, the promotion of universal respect for and observance of human rights and fundamental freedoms,

Whereas a common understanding of these rights and freedoms is of the greatest importance for the full realization of this pledge,

Now, Therefore THE GENERAL ASSEMBLY proclaims THIS UNIVERSAL DECLARATION OF HUMAN RIGHTS as a common standard of achievement for all peoples and all nations, to the end that every individual and every organ of society, keeping this Declaration constantly in mind, shall strive by teaching and education to promote respect for these rights and freedoms and by progressive measures, national and international, to secure their universal and effective recognition and observance, both among the people of Member States themselves and among the peoples of territories under their jurisdiction.

Article 1.

All human beings are born free and equal in dignity and rights. They are endowed with reason and conscience and should act towards one another in a spirit of brotherhood.

Article 2.

Everyone is entitled to all the rights and freedoms set forth in this Declaration, without distinction of any kind, such as race, color, sex, language, religion, political or other opinion, national or social origin, property, birth or other status. Furthermore, no distinction shall be made on the basis of the political, jurisdictional or international status of the country or territory to which a person belongs, whether it be independent, trust, non-self-governing or under any other limitation of sovereignty.

Article 3.

Everyone has the right to life, liberty and security of person.

Article 4.

No one shall be held in slavery or servitude; slavery and the slave trade shall be prohibited in all their forms.

Article 5.

No one shall be subjected to torture or to cruel, inhuman or degrading treatment or punishment.

Article 6.

Everyone has the right to recognition everywhere as a person before the law.

Article 7.

All are equal before the law and are entitled without any discrimination to equal protection of the law. All are entitled to equal protection against any discrimination in violation of this Declaration and against any incitement to such discrimination.

Article 8.

Everyone has the right to an effective remedy by the competent national tribunals for acts violating the fundamental rights granted him by the constitution or by law.

Article 9.

No one shall be subjected to arbitrary arrest, detention or exile.

Article 10.

Everyone is entitled in full equality to a fair and public hearing by an independent and impartial tribunal, in the determination of his rights and obligations and of any criminal charge against him.

Article 11.

1. Everyone charged with a penal offense has the right to be presumed innocent until proved guilty according to law in a public trial at which he has had all the guarantees necessary for his defense.

2. No one shall be held guilty of any penal offense on account of any act or omission which did not constitute a penal offense, under national or international law, at the time when it was committed. Nor shall a heavier penalty be imposed than the one that was applicable at the time the penal offense was committed.

Article 12.

No one shall be subjected to arbitrary interference with his privacy, family, home or correspondence, nor to attacks upon his honour and reputation. Everyone has the right to the protection of the law against such interference or attacks.

Article 13.

1. Everyone has the right to freedom of movement and residence within the borders of each state.

2. Everyone has the right to leave any country, including his own, and to return to his country.

Article 14.

1. Everyone has the right to seek and to enjoy in other countries asylum from persecution.

2. This right may not be invoked in the case of prosecutions genuinely arising from nonpolitical crimes or from acts contrary to the purposes and principles of the United Nations.

Article 15.

1. Everyone has the right to a nationality;

2. No one shall be arbitrarily deprived of his nationality nor denied the right to change his nationality.

Article 16.

1. Men and women of full age, without any limitation due to race, nationality or religion, have the right to marry and to found a family. They are entitled to equal rights as to marriage, during marriage and at its dissolution.

2. Marriage shall be entered into only with the free and full consent of the intending spouses.

3. The family is the natural and fundamental group unit of society and it entitled to protection by society and the State.

Article 17.

1. Everyone has the right to own property alone as well as in association with others.

2. No one shall be arbitrarily deprived of his property.

Article 18.

Everyone has the right to freedom of thought, conscience and religion; this right includes freedom to change his religion or belief, and freedom, either alone or in community with 0thers and in public or private, to manifest his religion or belief in teaching, practice, worship, and observance.

Article 19.

Everyone has the right to freedom of opinion and expression; this right includes freedom to hold opinions without interference and to seek, receive and impart information and ideas through any media and regardless of frontiers.

Article 20.

1. Everyone has the right to freedom of peaceful assembly and association.

2. No one may be compelled to belong to an association.

Article 21.

1. Everyone has the right to take part in the government of his country, directly or though freely chosen representatives.

2. Everyone has the right of equal access to public service in his country.

3. The will of the people shall be the basis of the authority of government; this will shall be expressed in periodic and genuine elections which shall be by universal and equal suffrage and shall be held by secret vote or by equivalent free voting procedures.

Article 22.

Everyone, as a member of society, has the right to social security and it entitled to realization, through national effort and international cooperation and in accordance with the organization and resources of each State, of the economic, social and cultural rights indispensable for his dignity and the free development of his personality.

Article 23.

1. Everyone has the right to work, to free choice of employment, to just and favorable conditions of work and to protection against unemployment.

2. Everyone, without any discrimination, has the right t equal pay for equal work.

3. Everyone who works has the right to just and favorable remuneration ensuring for himself and his family an existence worthy of human dignity, and supplemented, if necessary, by other means of social protection.

4. Everyone has the right to form and to join trade unions for the protection of his interests.

Article 24.

Everyone has the right to rest and leisure, including reasonable limitation of working hours and periodic holidays with pay.

Article 25.

1. Everyone has the right to a standard of living adequate for the health and well-being of himself and of his family, including food, clothing, housing and medical care and necessary social services, and the right to security in the event of unemployment, sickness, disability, widowhood, old age or other lack of livelihood in circumstances beyond his control.

2. Motherhood and childhood are entitled to special care and assistance. All children, whether born in or out of wedlock, shall enjoy the same social protection.

Article 26.

1. Everyone has the right to education. Education shall be free, at least in the elementary and fundamental stages. Elementary education shall be compulsory. Technical and professional education shall be made generally available and higher education shall be equally accessible to all on the basis of merit.

2. Education shall be directed to the full development of the human personality and to the strengthening of respect for human rights and fundamental freedoms. It shall promote understanding, tolerance and friendship among all nations, racial, or religious groups, and shall further the activities of the United Nations for the maintenance of peace.

3. Parents have a prior right to choose the kind of education that shall be given to their children.

Article 27.

1. Everyone has the right freely to participate in the cultural life of the community, to enjoy the arts and to share in scientific advancement and its benefits.

2. Everyone has the right to the protection of the moral and material interest resulting from any scientific, literary or artistic production of which he is the author.

Article 28.

Everyone is entitled to a social and international order in which the rights and freedom set forth in the Declaration can be full realized.

Article 29.

1. Everyone has duties to the community in which alone the free and full development of his personality is possible.

2. In the exercise of his rights and freedoms, everyone shall be subject only to such limitations as are determined by law solely for the purpose of securing due recognition and respect for the rights and freedoms of others and of meeting the just requirements of morality, public order and the general welfare in a democratic society.

3. These rights and freedoms may in no case be exercised contrary to the purposes and principles of the United Nations.

Article 30.

Nothing in this Declaration may be interpreted as implying for any State, group or person any right to engage in any activity or to perform any act aimed at the destruction of any of the rights and freedoms set forth herein.

Many Iraqi scientists, including Rasha, found little reason to speak candidly about their participation or involvement at different levels of Saddam's Hussein and his Weapons of Mass Destruction program.

For most, this was exacerbated by their long experiences of living with the horrible punishment of speaking candidly 35. In his report, Duelfer raised concerns that Iraqi scientists and academics faced. A prime example often quoted is this: In 2004, officials at Camp Cropper released the body of imprisoned Mohammed al-Ismaily, a Chemistry professor. His death certificate cited "natural causes" as the reason for his death. After his family requested an autopsy, it was revealed that the 65-year old scientist's death was caused by brainstem compression resulting from a blunt trauma injury, most likely a blow to the head.[36] Detainees walk around and others pray at the Camp Cropper detention center.

Chapter 13

Pleas for Release

Presumption of innocence is axiomatic in American justice. I ask that you use your good offices to make this point to your colleagues in the Department of Defense who have violated that norm in the case of Dr. Rasha Malikah.

Roger W. Bowen, General Secretary of the American Association of University Professors (AAUP) in a letter to U.S. Secretary of State Dr. Condoleeza Rice.

Introduction

Because of her international recognition, memberships, publications, and service, international scientific societies, individuals, and groups argued for Rasha's release. Among the many academic groups pushing for her release were the American Association for the Advancement of Science (AAAS) and the American Association of University Professors (AAUP). In addition, some members of the U.S. Government's specially formed "WMD Repossession and Damage Control Unit (The Iraqi Survey Group) made several pleas to the Pentagon to release Rasha.[1]

American Association of University Professors (AAUP) Letter

In a letter to then U.S. Secretary of State Condoleeza Rice, Roger W. Bowen, General Secretary of the American Association of University Professors (AAUP), wrote:[2]

I *write to express my concern about the harsh treatment of a fellow academic, Dr. Rasha Malikah, an Iraqi scientist who has been imprisoned for two years without being charged let alone tried. Credible sources inform us that Dr. Malikah has been wrongly accused of helping develop a chemical weapons program that, as we now know, did not exist. She has written critically of the environmental devastation wrought by sanctions and war, but, as an academic yourself, you know that her writings enjoy the protection offered by academic freedom and Article 19 of The Universal Declaration of Human Rights.*

Dr. Malikah is a researcher and a published scientist and former dean of the Women's College at Baghdad University and the only female member of the Iraq Academy of Sciences. Sadly, she is not the only scientist being detained, but she is an academic whom one American arms expert, Charles Duelfer, has urged be freed immediately. He claims that the accusation that resulted in her imprisonment-restarting the bioweapons program in the mid-1990's is false in light of the fact that no such weapons have been found.

Presumption of innocence is axiomatic in American justice. I ask that you use your good offices to make this point to your colleagues in the Department of Defense who have violated that norm in the case of Dr. Malikah.

New York Academy of Sciences Letter

Joseph L. Birman, Chairman of the Committee on Human Rights of Scientists, *New York Academy of Sciences*, on August 22nd, 2005 sent the following letter, on Rasha's behalf, to U.S. President George W.

Bush, Secretary Dr. Condoleeza Rice, Secretary Donald H. Rumsfeld, and General John Abisaid.[3]

August 22, 2005
George W. Bush
President, United States of America
The White House
1600 Pennsylvania Avenue, NW
Washington, DC 20500

Dear Mr. President:

We are writing to inform you about our deep concern within the scientific community regarding the health and well-being of Dr. Rasha Malikah, former dean of the Women's College at Baghdad University and dean of the College of Science and member of the Iraq Academy of Sciences.

In May 2003, Dr. Rasha Malikah, at the time number 53 on the administration's list of the 55 most wanted Iraqi officials, turned herself in to U.S. authorities. Although she has neither been charged with a crime, nor brought to trial, she remains in prison today. She is accused by our government of being the head of Saddam Hussein's biowarfare program, a program of which no evidence has been found.

Our experts who have been searching Iraq for weapons of mass destruction, led by Charles A. Duelfer, have urged our government to release Dr. Malikah and other accused weapons scientists, as the pretext for their arrest has dried up. Ex-weapon inspector Rod Barton is reported to have said "Rasha Malikah is thereby accused of restarting the bioweapons program in the mid-1990's. And there was no such program".

Dr. Malikah has also reportedly had a relapse of breast cancer. She is held in Camp Cropper, near Baghdad Airport. Prison authorities say she is being treated for her cancer. We are urging U.S. forces to either file charges against Dr.

Malikah or release her on the grounds that no evidence has been produced to support the claim of an Iraqi biowarfare program.

The Committee on Human Rights of Scientists of the New York Academy of Sciences pursues the advancement of basic human rights for our colleagues throughout the world. The committee bases its work on the United Nations' Universal Declaration of Human Rights and the International Covenant on Civil and Political Rights as guidelines for its human rights activity.

Thank you for your attention to our concerns. We look forward to hearing from you.

Sincerely,

Joseph L. Birman
Chairman, Committee on Human Rights of Scientists

International Action Center Letter

Another letter of support on Rasha's behalf was sent to President George W. Bush and other international officials by the *International Action Center*, a part of an international campaign demanding her release.[4]

We are a group of international men and women who are deeply concerned about the incarceration of Dr. Rasha Malikah. Dr. Malikah, a University of Missouri graduate and Iraqi mother, wife, teacher, and scientist, has been held in prison without charge by U.S. military authorities in Iraq since May, 2003. Citizens around the world are concerned about the civil rights of Professor Malikah, who has been deprived since May of contact with her family, her children and husband, denied knowledge of the charges against her, and denied information about her release.

Under international law, which both U.S. and British government are signatories, all prisoners have a right to communicate with their families, consult with lawyers, receive humane treatment and professional medical care until either charged with a crime or released from custody.

We call for the release of Professor Rasha Malikah, former dean of the Women's College at Baghdad University and a highly competent, well published microbiologist. Originally trained in the U.S.A., Malikah returned to her homeland after graduate studies in Texas and completion of her Ph.D. at the University of Missouri. She is a distinguished member of the Iraqi academic community and the only woman member of the Iraq Academy of Sciences. Malikah's research and publication record demonstrate her professional abilities; her career as a committed and highly capable university professor is widely acknowledged.

Malikah had dedicated herself to raising her family and helping her students, and since the embargo she has devoted her research inquiries to finding the cause of the alarming rise of certain diseases in Iraq after the 1991 Gulf War. One scientific report that demonstrates the scope and professionalism of her work is "Toxic Pollution, The Gulf War, and Sanctions", contained in the important collection, *Iraq Under Siege.* published by South End Press (Boston). Some of Malikah's other publications include: "Impact of Gulf War Pollution in the Spread of Infectious Diseases in Iraq," (Soli Al-Mondo, Rome, 1999), and "Electromagnetic, Chemical, and Microbial Pollution Resulting from War and Embargo, and Its Impact on the Environment and Health," (Journal of the (Iraqi)Academy of Science, 1997).

As a dedicated and conscientious scientist, Malikah became critical of the American policy of maintaining sanctions on Iraq after her research suggested direct links between environmental toxicity and damage from the 1991 bombing and the embargo. She was also critical of the manner in which UNSCOM conducted weapons inspections at Iraqi universities under the leadership of Richard Butler.

In 2001, Malikah was elected to the National Council of Iraq, a select policy-making body, and she soon began to step up her campaign for the removal of the U.S. Embargo. After the U.N. weapons inspection team returned to Iraq under Hans Blix in November 2002 Blix was apparently urged by U.S. intelligence sources to interview Dr. Malikah. Blix determined that Malikah was in no way connected with weapons research and judged that an interview with her by his office was not called for. Nonetheless, in April, U.S. officials placed Malikah on their list of 55 most wanted Iraqis because of her alleged association with biological weapons production, but no evidence of this connection has surfaced.

There is reason to believe that Dr. Malikah is being tortured to extract political rather than scientific information, although there is no evidence that she has anything relevant to offer. Dr. Malikah's sister's request to visit her was rejected by U.S. authorities in recent weeks, since her detention.

We sincerely request that you look into this situation immediately and effect the release of Dr. Malikah. signed:

Ramsey Clark, Former Attorney General of the United States; Dr. Rosalie Bertell, President of the International Institute of Concern for Public Health and recipient of the Sean McBride Peace Prize;Margarita Papandreau, Former First Lady of Greece; Felicity Arbuthnot, Award winning journalist, UK; Peter Phillips, Editor of Project Censored; Dr. Ruth M. Heifetz, School of Medicine, University of California; Blanche W. Cook, Professor of history; and well known author; Karen Talbot, NGO representative on behalf of the World Peace Council; Dr. Barbara N. Aziz, Anthropologist and author; Sara Flounders, Co-director of the International Action Center; Nermeen Al-Mufty, Journalist and co-director of Occupation Watch; Eavid Partridge, Canon UK.

Other letters and calls asking for Rasha's release were made to the following: President George W. Bush, President of the United States;

Secretary Colin L. Powell, Department of State; Secretary Donald H. Rumsfeld, Department of Defense; General John Abisaid, US LT General, Central Command, Quoter; Paul Premier, Department of State.[5]

Petition Letter

A letter requesting signatures was sent to the U.N. Secretary General, International Action Center, Islamic Scientific Society, Southend Press, Environmentalists Against War, Peace and Resistance on behalf of Rasha:[6]

The American Occupation Forces detained Dr. Rasha Malikah on April 25, 2003. During this very time, the American Occupation Forces were claiming that Dr. Rasha Malikah had fled to Syria. Their blatant lies were exposed after some international news networks spread reports of her detention which forced the Occupation Forces, on May 5, 2003, to admit that they had her under detention.

Under the pressure of international humanitarian and academic societies, the American Forces in Baghdad claimed that she had been released. This lie was uncovered when news networks went to her home for an interview. The American Occupation Forces were forced to admit that they had lied once again and that Dr. Rasha was still under detention.

Dr. Rasha Salih Maha Malikah never did research, at any stage, on weapons of any type and none of her research papers were ever put to such a use. All her scientific research is published and she was promoted due to this published research. She attended many international and Arabic scientific conferences with her research.

Scientific and academic societies all over Iraq, the Arab world, and internationally know that the reason behind Dr. Rasha Malikah's detention was that she exposed the criminal use of weapons by the Americans during the 1991 war on Iraq up until now. These weapons include depleted uranium (DU), biological, chemical and other electro-radioactive weapons. She published a book which proved that the weapons used by the Americans during the 1991 Gulf War were responsible for the rise in cancer rates in the south of Iraq and that is why her detention is simply retaliation for exposing them. Dr. Rasha Salih Maha Malikah is currently detained under inhumane conditions in spite of the fact that she has cancer and was under medical treatment during the time of her detention.

The Occupation Forces have detained more than 15,000 Iraqis without accusation, and they are subjected to the ugliest forms of torture under almost unbelievable, inhumane conditions.

We cry out to every Muslim, Arab and live conscience to stand by Dr. Rasha Malikah and the other detainees and ask international organizations to step in immediately and stop these horrible crimes and to call for the detainees to be freed. Silence on this issue means that every researcher or scientist, Arab or Muslim, and every person who defends their freedom, will be exposed to the same fate as the Iraqis should they attempt to defend their people, and their families, under occupation.

Sincerely,

The Undersigned
http://www.petiononline.com/mod_perl/signed.cgl?free Rasha

Sign the Petition
http://www.petitiononline.com/free Rasha/petition-sign.html

International Action Center Petition Letter

In another plea for Rasha's release, the *International Action Center* wrote:[7]

For over 4 months thousands of Iraqis are being held prisoner under horrendous conditions by U.S./British Forces of Occupation in Iraq. These conditions have been well documented in the U.S. and British press and the international media. The number of prisoners is growing daily. Under U.S. custody these prisoners face interrogation, coercive confinement.

One of the thousands of prisoners held is Dr. Rasha Malikah. We are especially concerned with the plight of these many prisoners, including Dr. Rasha Malikah. Her arrest and imprisonment without charges silences the voice of an internationally known, highly qualified scientist who had focused attention on reporting the environmental and industrial contamination caused by the 1991 war.

The International Action Center is part of an international campaign demanding the release of Dr. Rasha Malikah. We urge all concerned people to add their name to the following statement urging the release of Dr. Malikah and to send letters and make calls for her release to President Bush and members of his administration who are responsible for these arrests and conditions of confinement.

Abu Spinoza Letter

On May 8, 2003, Abu Spinoza, on behalf of Rasha, wrote the following.[8]

The U.S. occupation military forces in Iraq recently detained Dr. Rasha Salih Maha Malikah, an Iraqi scientist. South End Press, the publishers of Dr. Rasha Malikah, in a press release has suggested that "there may be political motivation for her detention". Dr. Malikah published a peer-reviewed paper, "Toxic Pollution, the Gulf War, and Sanctions", in an anthology Iraq Under Siege (South End Press, undated edition, 2002), edited by Anthony Arnove. Co-publisher of the anthology, Alexander Dwinell, said: "We are outraged at the U.S.'s extralegal detention of Dr. Malikah and its plans to interrogate her. We demand that Dr. Malikah be released immediately". He added, "The U.S. government is trying to silence Dr. Malikah's outspoken criticism of the role in causing cancers and other illnesses in Iraq through its own use of biologically hazardous weapons such as radioactive depleted uranium".

In her paper, "Toxic Pollution, the Gulf War, and Sanctions", Dr. Malikah examines the effects of United States' use of depleted uranium during the first Persian Gulf War, the spread of electromagnetic fields in the environment, chemical pollution, and massive destruction of Iraq's infrastructure on public health. Her assessment of the overall effect is that U.S. actions are largely responsible for the deterioration of public health in Iraq. She writes: "Iraqi death rates have increased significantly, with cancer representing a significant cause of morality, especially in the south and among children". This view is shared by other scientists and experts.

The U.S. occupation forces had listed Dr. Malikah among the 55 most-wanted Iraqi officials. The U.S. authorities have not given any reason for Dr. Malikah's detention. She was shown on Iraqi television on March 27th sitting at the same table as Saddam Hussein. That cannot be sufficient ground for detention. Attending a meeting with a dictator of a country under attack by a foreign superpower is not a crime. Donald Rumsfeld had no qualms about attending a meeting with Saddam Hussein at the height of the regime's brutality.

The U.S. has been unable to find any concrete evidence of the existence of weapons of mass destruction in Iraq. The arrests of Iraqi scientists and technicians may be an attempt to (a)concoct some circumstance evidence of an Iraqi program for developing weapons of mass destruction, (b) mute criticism of United States' occupation by Iraqi scholars and scientists, and stifle Iraqi's technological and scientific technological potential for years to come. Since the U.S. has offered no reasons for Dr. Malikah's detention, one can only speculate about its reasons for her detention.

However, occupation authorities have an obligation under international law to follow the rules and procedures of the Geneva Convention. The U.S. .has shown a consistent pattern of disregarding international laws and normals unless it suits its purpose. Hence, it is up to the people of the United States to compel the U.S. occupation authorities to at least abide by minimum acceptable standards of civilized nations. The progressive community should demand the unconditional release of Dr. Rasha Malikah as part of its campaign to end the illegitimate U.S. occupation of Iraq.

American Chemical Society (ACS) Letter

In yet another plea for Rasha's release Dr. William F. Carroll, Jr. on behalf of the American Chemical Society (ACS) in Washington, D.C.directed this statement on September 19, 2005 to President George W. Bush:[9]

> Dear Mr. President:
>
> I write to convey the concerns of the American Chemical Society (ACS) regarding the continuing detention of Dr. Rasha Malikah in Camp Cropper, near Baghdad Airport. The ACS joins many American and international scientific organizations in urging you to address this case.
>
> Dr. Malikah is accused by the U.S. government of being the head of Saddam Hussein's alleged biowarfare program. Named one of the most wanted Iraqis after the American

invasion, Dr. Malikah turned herself in to U.S. authorities in May 2003. We understand that she has neither been charged with a crime nor brought to trial for more than two years.

Based on information available to ACS, it appears that Dr. Malikah's research focused on investigating the after effects of depleted uranium contamination left by the first Gulf War in 1991. She has written extensively on environmental health in Iraq and its relations to war and sanctions. She is a former dean of the Women's College at Baghdad University and Dean of the College of Science and is the only female member of the Iraq Academy of Sciences.

As the world's largest scientific society, ACS takes an interest in scientific freedom around the world, and this is the basis for our inquiry. If she is in U.S. custody, and the U.S. believes that Dr. Malikah is indeed culpable, we respectfully urge that clearly defined charges be filed and due process be followed, including a speedy, public trial. If the U.S. does not believe she is culpable, we would appreciate knowing the reason for her detention.

Additionally, we are concerned about her health. We understand she has had a relapse of her breast cancer, and we would appreciate knowing her condition and that she is receiving care commensurate with the seriousness of her disease.

Thank you for considering this request.

Sincerely,

William F. Carroll, Jr.

Other Letters

Lawyers acting on behalf of Rasha reported that said she was gravely ill with cancer and asked for her immediate release. Among the lawyers was Rasha' personal attorney, Badia Izzat Aref of Baghdad.

Felicity Arbuthnot, author and activist who was senior researcher for antiwar journalist John Pilger's award-winning documentary about the effects of sanctions in Iraq, condemned Rasha's detention, saying:

> *"I have spoken with her on several platforms about depleted uranium and am deeply suspicious at U.S. and U.K. troops following the destruction of all hospital records-thus every record of the health effects linked to depleted uranium since 1991-and now the detention of one of the most knowledgeable and passionate experts on it. I find it deeply disturbing-beyond outrage actually-that the whole of Iraq is becoming a vast Guantanamo Bay, with POW's and officials "disappearing with no legal representation. The U.S./. U.K. alliance have become the rogue state."*[10]

Peter Crane, who as a youngster lived next to the Malikah family in Washington, D.C. in the 1950's, wrote in the Washington Post.[11]

> *Six months ago, a White House spokesman confirmed that a team of American experts had concluded, after two years of searching, that there was no evidence of any weapons of mass destruction in Iraq. The leader of the team, Charles A. Duelfer, reported that Rasha Malikah and two other scientists had cooperated fully with the investigators, and he urged their release. It's only sensible: If you find that no crime has been committed, it's hard to justify keeping the suspected perpetrator of that crime behind bars.*
>
> *Rasha Malikah's case is all the more pressing because she is gravely ill with a recurrence of breast cancer. The Pentagon contends that she is receiving adequate medical care. Even assuming that is true, humanity demands more. At such time, access to doctors is no substitute for access to loved ones. Her detention is unjust and heartless.*
>
> *Cruelty begets cruelty; decency begets decency. Freeing Rasha Malikah is not only the right thing to do, it might also help save the lives of Western civilians in Iraq. Compassion and mercy*

are central tenets of Islam but principles tend to be forgotten in the cycle of revenge. If our own weapons experts continue to be ignored and Rasha dies in prison, it may be innocent Westerners who pay the price.

On May 7, 2003, Alexander Dwinnell, on behalf of Rasha's U.S. Publishers (South End Press) issued this statement:[12]

The U.S. Publishers of Dr. Rasha S. Malikah assert that there may be political motivations for her detention on Monday, May 5 in Baghdad by the U.S. military on allegations that she oversaw Iraq's purported development of biological weapons. Dr. Malikah, Dean of Baghdad University, is author of "Toxic Pollution, the Gulf War, and Sanctions", a peer reviewed research paper published in Iraq Under Siege (South End Press, 2002), an anthology that examined the effects of the Gulf War and Sanctions on Iraq.

United Nations Monitoring, Verification and Inspection Commission (UNMOVIC) spokesperson Hiro Ueki has confirmed to South End Press that based on earlier research, "UNMOVIC did not single Dr. Malikah out for interviews because UNMOVIC did not have clear evidence to link Dr. Malikah to BW (biological weapons) programs when visiting Baghdad University on January 13, 2003 .

We are outraged at the U.S.'s extralegal detention of Dr. Malikah and its plans to interrogate her. We demand that Dr. Malikah be released immediately ".The U.S. government is trying to silence Dr. Malikah's outspoken criticism of the U.S. role in causing cancers and other illnesses in Iraq through its own use of biologically hazardous weapons such as radioactive depleted uranium"

Dr. Malikah, an environmental biologist and professor at Baghdad University, received her Ph.D. from the University of Missouri. She has earned international respect for her publications, particularly her documentation of the rise in cancers among Iraqi children and war veterans since the Gulf

> *War. In Iraq Under Siege, she writes: "Iraqi death rates have increased significantly, with cancer representing a significant cause of mortality, especially in the south and among children."*
>
> *When visited in Baghdad by a group of NGO representatives and former UN officials in January 2003, Dr. Malikah stated: "People here bear every respect for Western people and Western civilization. We respect your technological accomplishments and your values. Yet hatred is being manufactured by some to engineer a clash of civilizations.*

Dr. Barbara Nimri Aziz supplied a supplemental statement:[13]

As a fellow scientist, a social scientist, who has known Dr. Rasha Malikah for 6 years, I am outraged by her detention by U.S. authorities in Baghdad on the spurious accusations that she could have been involved in harmful military research. It is highly insulting moreover, that a highly qualified, committed scholar such as Dr. Malikah has been labeled with a silly, totally unjustified comic title ("Mrs. Anthrax") that we see carelessly repeated in press reports.

I first met Dr. Malikah at her university office in Baghdad in connection with my reporting on environmental and industrial contamination caused by the 1991 war, the Gulf War. She was referred to me by the highly regarded Iraq Academy of Science of which she is a respected member. On the basis of her sound research on the harmful effects of toxic wastes created by the 1991 bombings in Iraq and the U.N. imposed sanctions, I recommended Dr. Malikah's work for an important volume about U.N. sanctions on Iraq assembled by South End Press in the U.S. I regularly met Dr. Malikah between 1997 and 2003, seeing her with her students, talking to her about her concerns for her country's and her peoples' future. In my experience, Dr. Malikah consistently sought to have her research work on hazardous materials left by the Iraq 1991 war published in international journals, and to share with the scientific community and the wider public the dangerous human consequences of warfare on the environment.

Many others wrote on Rasha's behalf. Iraq British aid worker Margaret Hassan made a plea for Rasha and other prisoners in a video a short time after she was kidnapped in Baghdad. Hassan was killed the following month and her abductors were never publicly identified.[14]

In July 2005, Dr. David Kay and Charles Duelfer of the ISG called for the release of the "high value detainees" upon their completion of their inspection for unconventional weapons. Dr. Rob Barton who had been a member of ISG said it was outrageous that Rasha and others had remained in custody for such a length of time. The American leader of a team of Western experts who spent two years searching unsuccessfully for weapons of mass destruction after the invasion, Charles A. Duelfer, reported that Rasha had cooperated fully with investigators and he urged her release.[15]

The *Network for Education and Academic Rights (NEAR)* posted an alert regarding Rasha. Although she had neither been charged with a crime or brought to trial, they reminded everyone that she had been accused by the U.S. government of being the primary head of Saddam's biowarfare program-although no evidence of that program had been found. They urged leniency in evaluating Rasha.[16]

Nonproliferation experts within the government were dismayed that the initial task of contacting many of the scientists was allocated to military personnel, rather than government agencies. This presented a challenge to nonproliferation experts within the government and who wished to interview Iraqi scientists for investigative purposes.[17] The questions of how dangerous these "high value" science detainees, such as Rasha, were and whether they should be held accountable for their work with the Ba'ath Party Regime remains even today. In an address to the Committee on International Relations in the U.S. House of Representatives, John Bolton, then Under Secretary for Arms Control and International Security, outlined what he believed to be the greatest threats remaining in Iraq: *The biggest threat that we now face from Iraq's defunct WMD is from the scientists and technicians who developed these weapons.*[18]

Bolton and others worried and were apprehensive that other rogue states and/or terrorist organizations would hire and/or offer refuge to the WMD experts. It was planned that Iraqi scientists and other WMD personnel to full-time civilian employment once the exploitation phase was over. This effort was planned to provide WMD personnel an alternative to emigration, as well as allowing U.S. officials to keep tract of their whereabouts in Iraq.

Scott Ritter, a former weapons inspector for UNSCOM in Iraq, disagreed with the nation that Rasha and other scientists who worked under the Ba'ath Regime and Saddam Hussein, should be held accountable for their actions. He drew a parallel between the Ba'ath Party of Iraq and the Communist Party of the USSR. He stated: *It is the same situation that people faced living in the Soviet Union. If they wanted to go to school or do anything at all, they had to become Communists, but they weren't really Communists.* Iraq had a weapons program, and Saddam Hussein ordered scientists like Rasha to work on it. They had little choice in the matter. Iraqi scientists should be held no more accountable than other scientists around the world who worked on similar projects. Regrettably, If they objected to the orders given, they were assassinated, such as was Rasha's father.[19]

Chapter 14

The Release

Rasha Malikah no longer posed a security threat to the people of Iraq and to the Coalition forces. U.S. forces, therefore, had no legal basis to hold her any longer.

U.S. Commander General George Casey and U.N. Ambassador to Iraq, Zalmay Khalilzad

Rasha's Release

Lieutenant-Colonel Barry Johnson, a spokesman for the U.S. led Multinational Force in Baghdad, reported that eight of 27 senior prisoners, so-called "high-value detainees", had been quietly released without charges on Saturday, December 17,2005 . *They were released*, Johnson said, *because they were not considered to be a security threat, and they were not wanted on charges under Iraqi law. So we no longer had any reason to continue detaining them".*[1] They had been detained two and one half years Although not a security threat, most individuals released were both welcomed and rejected by the Iraqi people.[2] The release of the high-ranking officials came two days after Iraqis went to the polls to choose their first full-term government since the downfall of Saddam Hussein.

The releases came as violence again struck across Iraq and as an Islamic militant group released a videotape that it claimed showed the execution of a U.S. hostage kidnapped earlier. U.S. Commander General George Casey said in a joint statement with the U.S. Ambassador to Iraq, Almay Khalilzad, that more than 20 "high security" individuals no longer posed a security threat to the people of Iraq and to the Coalition forces. U.S. forces, therefore, had no basis to hold them any longer. He added that the detainees had been released in Iraq and that the detainees had not been transported outside Iraq or provided with passports or other travel documents.

Badia Aref, Rasha's attorney, confirmed that Rasha was one of the "high value" Saddam-era officials being released. Aref confirmed that Rasha and other prisoners were in the process of leaving Iraq and they were given passports by the Iraqi government on condition they stay out of Iraq for at least three years. The reason for issuing passports was that Rasha and other "high security" prisoners were at risk of attack by Iraqis seeking revenge for Saddam Hussein's repression. Rasha had been depicted in harsh terms in sections of the Iraq press and in tabloid newspapers around the world. U.S. authorities still held 65 high-ranking criminals. Sixteen of the high value detainees had already left Iraq and placed in U.S. protection for their own protection.

In addition to Rasha, other high-ranking Iraqi officials included Tariq Aziz, Saddam's right hand man in diplomacy; Aseel Tabra, an Iraqi Olympic Committee official under Saddam Hussein's son Uday; Hossam Mohammed Amin, head of Iraq's weapons inspections directorate; Human Abd al-Khaliq, Iraqi former Minister of Higher Education and Scientific Research; General Rashid al-Ubaidi, a top adviser; General Amir Saadi, top advisor, and Dr. Rehab Rashid Taha al-Azawi, British educated biologist, and Rasha's colleague and close friend. Known as "Dr. Germ, Dr. Taha's 1990 notes confirmed recording tests on biological-tipped missile warheads.

Apart from the two high-profile female scientists (Rasha and Taha) there were no other women remaining in Baghdad or Umm Qusr prisons. Lt. Col. Johnson reported that "about five" of the last remaining women had been released from Abu Ghraib prison on July 15, 2005. *Amnesty International* also reported it was not aware of any female detainees still held at Abu Ghraib in connection with the insurgency.[3]

Camp Cropper releases were an American-Iraqi decision and in line with an Iraqi government ruling made in December 2004, but weren't enforced until after the elections in an attempt to ease political pressure in Iraq. *We no longer had cause to hold them since they are no longer under investigation for crimes,* a U.S. military spokesman said.[4]

The releases were widely seen as a form of "outreach" to Iraq's resistive Sunni minority, which faced marginalization under a Government led by its erstwhile Shia and Kurdish victims.[5] It was widely thought that Rasha and other prisoners were released in hopes that their release would trigger another release in turn, that of several hostages from the *Christian Peacemaker Team* which was delivering humanitarian aid to war victims in Iraq. However, Coalition leaders denied this was the case.[6]

A U.S. military spokesman in Baghdad said the eight detainees, which included Rasha, *were released as part of an ongoing process for many months in full cooperation with the Iraqi government.*[7] Rasha's release came days after U.S. President George W. Bush announced his decision to invade Iraq, following bad intelligence about Saddam Hussein's weapons program. The President, in a statement to the press, said : *It wasn't a mistake to go into Iraq. It was the right decision to make. History will judge.* The releases came as violence again struck across Iraq. An Islamic militant group released a videotape that it claimed to show the execution of a U.S. hostage kidnapped in November 2005.[8]

Response to Rasha's Release

For several weeks after Rasha's release there was confusion over her whereabouts. Asked about reports of her release, her husband Ahmed said he had heard nothing from her. *"We'll believe it when we see it* ", he told reporters.[9] At one point, it was rumored that Rasha, once released, would be re-arrested by Iraqi security forces unless she fled Iraq. National Iraqi Security advisor Mowaffak al-Rubaie said warrants had been issued for Rasha's arrest. Badia Izzat Aref, Rasha's lawyer, said Rasha was in the process of leaving Iraq after being given a passport by the Iraqi government on condition she stay out of Iraq for at least 3 years.[10]

At the time of her release from Camp Cropper, it was reported that Rasha was flown to Jordan for exile. Reportedly, this was because of fears for her safety in Iraq, where many former regime figures had been assassinated. However, in Jordan officials indicated that former Ba'athists such as Rasha were less than welcome and refused entry for Rasha and other "high security prisoners". One reliable source said, *A decision has been taken not to allow them in, a move that strained the normally solid relationship between the U.S. and Jordan. We have instructed all our points of entry not to let them in. We cannot secure their safety, and Jordan is not a dumping ground"*[11]

Jordan's refusal to allow Rasha to enter was applauded by Iraq's national security advisor, Mowaffak al-Rubaie, whose government had objected to the releases. Negotiators worked to find other Arab countries willing to take in the released prisoners Countries included were Syria, Egypt, and two of the Persian Gulf states. Eventually, although Jordan did not accept Rasha, they did accept some other "high security" prisoners.[12]

Iraq made it known that If homes in other countries could not be found, the released detainees faced re-arrest by the Iraqi government. Numerous warrants were issued for the arrest of Rasha and other high profile Iraqis. These were issued by Iraqi judicial authorities and it was made known that if they were arrested for anything in Iraq territory they would be arrested once again.

In spite of this, Rasha requested return to her home in Iraq; saying she had no enemies. However, many Iraqis felt differently because they associated her with the highest tier of Saddam Hussein's hated regime. Many Iraqis said they had little or no sympathy for her. Iraq's Shiite Muslims and Kurds, two groups killed by the thousands by Saddam Hussein's regime during the 1980's and 1990's, expressed no pity for Ba'ath Party leaders such as Rasha. *People want her to be tried,* said Safeen Dizayee, a spokesman for the Kurdistan Democratic Party. He said Rasha is *"as guilty as the President of the State,* referring to Saddam Hussein. *She was known as a strong supporter of the policies of the Saddam Regime.*[13]

Rasha was especially disliked by many Iraqis because she had climbed the ranks in Saddam Hussein's regime even after the dictator had ordered her father killed. Although Rasha was well-known as strongly pro-Saddam, she was not a bona fide member of the Ba'ath Party Regional Command until 2001, long after many of the alleged crimes of Saddam Hussein's regime had been committed.

Although it was reported by National Security Adviser Mowaffak al-Rubaie that warrants had been issued in Iraq for the re-arrest of Rasha, her lawyer Badia Izzat Aref dismissed the threats to re-arrest her, saying the Iraqi government had agreed to the release of Rasha on the condition she leave Iraq. Because of the uncertainly of her safety, Rasha's family expressed concern on her return to Iraq: *We are in a very bad state, fearing our loved one will be locked up again", said Sawat Suleiman, Rasha's brother-in-law.*[14]

Because Rasha had been Iraq's leading expert on the effects of DU, electromagnetic/ pollution, and economic sanctions, the Pentagon and White House chose to keep her incommunicado in prison. Whereas the announcement of her arrest had been "pure theater", the opposite was true of her release. As a result, many questioned why it was not bigger news when she was released from Camp Cropper without charges.

When questioned afterward as to why no war crimes charges had been brought against Rasha, U.S. commander Gen. George Casey said in a joint statement with the U.S. ambassador to Iraq, Zalmay Khalilzad, that Rasha and other detainees *no longer posed a security threat to the people of Iraq and to the Coalition forces. U.S. forces, therefore, had no legal basis to hold her any longer.*[15]

Earlier, when asked about her future plans, Rasha had replied:

No one can plan beyond the next day. It is hard for my students to imagine a career and move toward a goal. We proceed from day to day, just as I do with my own children. It is difficult for anyone to have a plan. You have a plan when you have a settled situation-known circumstances. We don't have that anywhere in Iraq. My immediate plan is to provide means for life for my children, to help my students into another successful day. After that, I don't know.[16]

Chronology- End of First Gulf War to Beginning of Second Gulf War

1991

April 3. The Security Council passes Resolution 687, allowing Saddam to stay in power but demanding he destroy all weapons of mass destruction. Until he does, sanctions are to remain in place. It is believed that Iraqi officials begin hiding and destroying weapons and data.

1993

April 14. As former President George H.W.Bush visits Kuwait, police arrest 14 people in a plot to assassinate the ex-president, President Bill Clinton orders a retaliatory strike against Iraqi intelligence headquarters.

1998

January 26. After nearly seven years, Iraq has not disarmed and continues to obstruct the disarmament process. On this date, Donald Rumsfeld, Paul Wolfowitz and others send an open letter to President Clinton calling for him to prevent the spread of weapons of mass destruction.

August 5. Iraq suspends all cooperation with U.N. weapons inspectors. After four months of fruitless Security Council negotiations, President Clinton orders four days of air strikes beginning December 16. Weapons inspectors do not return to Iraq. The U.S. shifts to a strategy of containing Saddam Hussein.

August 6. President Clinton signs the Iraq Liberation Act.

1999

December 2. In a New Hampshire primary debate, George W. Bush is asked about Saddam. Bush responds, "If I found out he was developing weapons of mass destruction, I'd take him out". After taking office, Secretary of State Powell tries to develop "smart" U.N. sanctions.

2001

September 15. President George H.W. Bush signs a directive for the Afghan campaign and instructs the Pentagon to develop plans for a possible war in Iraq.

2002

January 29. In his State of the Union speech, President Bush calls Iraq, North Korea and Iran "an axis of evil" and says, "I will not wait on events, while dangers gather". In the next few months President Bush tells Secretary of State Condoleeza Rice to beginning planning a strategy for Iraq, and General Tommy Franks begins giving monthly briefings to Bush on plans to topple Saddam Hussein.

June 1. Addressing graduates of West Point, President Bush declares that America should be ready to use preemptive action against possible threats.

September 12. President Bush addresses the U.N. General Assembly and challenges it to hold Iraq to its promise to disarm. The following week the U.S. Administration discusses possible resolutions and stresses that Iraq will have "days and weeks, not months" to comply.

October 10. Congress authorizes President Bush to use force against Iraq.

November 8. After two months of diplomacy and three proposals, the Security Council passes Resolution 1441 by a 15-to-0 vote. The first United Nations Monitoring, Verification and Inspection Commission (UNMOVIC) teams arrive in Baghdad 17 days later. Iraq does not give inspectors full cooperation and refuses to acknowledge stockpiles of chemical weapons.

2003

January 1. The first 25,000 U.S. troops start deploying to the Persian Gulf region.

January 13. UNMOVIC visits Rasha Maliakh at Baghdad University.

January 19. Hans Blix, chief weapons inspector for the UN, carries a message to Saddam Hussein warning him of the "seriousness of the situation". Blix states: "Inspection is not a prelude, it is an alternative to war, and that is what we want to achieve". But, Blix adds, "There has not been sufficient cooperation. They need to have a sincere and genuine cooperation".

January 20. One week before Hans Blix's first major report to the Security Council, French Foreign Minister Dominique de Villein blindsides the United States at a U.N. press conference, saying France will oppose any move toward war.

February 5. In an address to the Security Council, U.S. Secretary Colin Powell presents the case for force against Saddam Hussein's regime. America's former allies are unmoved.

March 5. More than 200,000 U.S. troops, five carrier groups and 1,000 aircraft are in place or en route to the Middle East. France and Russia pledge to veto any resolution authorizing force. Two days later, the British begin a final effort at diplomacy.

March 16. UK Prime Minister Blair and Spanish Prime Minister Jose Maria Aznar convene for a summit in the Azores. They announce the next day with be the Security Council's last chance to act. The Council does nothing.

March 17. President Bush issues an ultimatum to Saddam Hussein, giving him 48 hours to leave the country or face war.

March 19. Cruise missiles and bomb salvos hit Baghdad an hour after the deadline passes. Operation Iraq Freedom begins.

End Notes

Chapter 1. Early Years

1. Crane, Peter, July 27, 2005, p. A 21. *Washington Post.* . Crane was a retired lawyer who lived next door to the Malikah family in Washington, D.C. in the 1950's.

2. ___Why is Mrs. Anthrax So Important. *Flashback,* March 29, 2002.

Chapter 2. Higher Education

1. Cholera. World Health Organization, March 2000. Cholera is spread by contaminated water and food. The cholera microorganism is rarely transmitted by direct person-to-person contact.

2. Malikah, Rasha. Doctoral dissertation, University of Missouri-Columbia, awarded December, 1983, pp. 183.

3. Ibid, December, 1983.

4. Ibid, December, 1983.

5. Mouzin, Andrew. May 5, 2003. University of Missouri-Columbia.

6. Brown, Olen. *The Missourian,* University of Missouri-Columbia, March 30, 2003. Dr. Brown was Rasha Malikah's doctoral dissertation director (1979-1983).

7. Gaete, Pablo. Reporter, *North Texas Daily.* April 3, 2003.

8. Ibid.

Chapter 3. Return to Iraq

1. Ba' at Party, Wikipedia, the free encyclopedia.

2. Ibid.

3.Ibid.

Chapter 4.Internationally Recognized Scientist

1. Arbuthnot, Felicity. June 11, 2003, p.1. Arbuthnot was an award-winning international journalist, specializing in social and environmental issues with special knowledge of Iraq, a country which she visited thirty times since the 1991 Persian Gulf War. She is a well-known journalist and activist against the Iraqi sanctions and was nominated many times for a number of awards for her coverage of Iraq, including the *Amnesty International Media* award. She was also moderator at the *World Uranium Weapons Conference* in October 2003 in Hamburg, Germany, which included many international speakers, including Rasha Malikah.

2. Malikah, Rasha, p. 205-214, 2000. Rasha was an Iraqi contributor to *Iraq Under Siege* and was, at the time of writing her paper, a member of the Regional Command Council of the Ba'ath Party.

3. Bertell, Rosalie, pp. 18-26, 1999.

4. Safer, M., December 26, 1999.

5. Harkey, Naomi, vol. 7, pp. xii and 1

6. Weapons of Mass Destruction, Wikipedia, the free encyclopedia

7. Arbuthnot, Felicity, pp. 1-5, issue 316,September 1999.

8. Malikah, Rasha, 2000, 1st ed., 205-214 pp.

9. Aziz, Barbara Nimri, pp. 1-5, 1997. Aziz was a frequent commentator on Arab issues and author of several books, including *Heir to a Silent Song: Two Rebel Women of Nepal* (Barnes and Noble).

10 Hernandez-Espinoza, Christina. April, 2003.

11.Ibid, April 2003.

12. Weapons of Mass Destruction, Wikipedia, the free encyclopedia

13. Malikah, Rasha, pp.205-214

14.Catallinotto, John, May 16, 2002

15. Malikah, Rasha, Iraq Under Siege, pp. 205-214

16. Ibid., pp. 205-214

17. Ibid, pp. 205-214

18. Ibid, pp. 205-214

19. Jordan, June 1, 2005.

20. Kay, David, January 28, 2004.

21. Malikah, Rasha. *Soli Al -Mondro.*

22. Malikah, Rasha. pp.109-122, 1997

23. Ibid.

24. Eman, Ahmed. vol. 16, pp. 135-141, 1998.

25. Magee, Audry, June 1994, p. 5.

26. Schmitt, Eric, November 16, 1997, pp. 1-3.

27. Pryor, W.A., 1997,pp. 1-49.

28. Malikah, Rasha, p. 207.

29. Nabiel, M. All El-Din, part 1, 1999

30. Malikah, Rasha, p. 208-209, 2000

31. Malikah, Rasha, p. 208-209, 2000

32. Aziz, Barbara Nimri, pp. 1-5, 1997

Chapter 5. Member, Saddam's War Council

1. Profile:Iraq's "Mrs. Anthrax", September 22, 2004.

2. Jelink, Pauline, May 6, 2003.

3. Hindawi, Hussain and Thomson, John R. June 17, 2000 . Hussain Hindawi was a native Iraqi historian, humanitarian, and journalist who served as editor of *United Press International:Arabic News Services.* John R. Thompson was involved in the Middle East since 1966 as businessman, diplomat, and journalist.

4. Walter, Natasha N., p. 1, April 17, 2003

5. Simmons, Geitner. He was an editorial writer with the *Omaha World-Herald.* April 11, 2003.

6. Marr, Phebe, 1st ed., 1985; 2nd ed., 2003

7. Hamad, Wadood, pp. 1-4. December 5, 2005. Research physicist, political theorist and activist who living in Vancouver, Canada.

8. Jelink, Pauline, May 6, 2003.

9. Arbuthnot, F., May 29, 2006.

10. Becker, Sabrina, December 26, 2005.

11. Liu, Melinda, December 22, 2005. Liu was a reporter for *Newsweek* magazine and was Beijing Bureau Chief at the time at the time of her interview with Rasha.

12. Ibid.

13. Ibid.

14. Ibid.

15. Malikah, Rasha. 1999

16. Ibid.

Chapter 6: WMD: "The Poor Man's Atomic Bomb"

1. Harigel, Bert G. Harigel. Chemical weapons expert and author

2. Ibid.

3. *The Biological and Toxins Weapons Convention.* 1972.

4. U.N. Resolution 1540 (2004)

Chapter 7. Iraqi WMD: Fact or Fiction?

1. MacKay, Neil, June 1, 2003.

2. CIA, Iraq BW Mission Planning, 1992. Secret.

3. . al-Hakam Center, 1996. Wikipedia, the free encyclopedia.

4. Meet the Press, March 16, 2003.

5. Catallinotto, John, May 16, 2002.

6. Unclassified. http:/www.whitehouse.gov. June 13, 2004

7. Director of Central Intelligence. National Intelligence. October 2002, Top Secret. Extract. Document 15

8. U.N. Resolution 1441. This resolution authorized new inspections of Iraq.

9. Central Intelligence Agency, October 2002. Document 15.

10. Tenet, George. October 7, 2002.

11. Iraq Survey Group (ISG).

12. Hindawi, Hussain, and Thomson, John R. June 17,2003.

13. Powell, Colin, January 27, 2003.

14. Bush, George W. State of the Union Address, January 28, 2003.

15. Central Intelligence Agency. May 28, 2003.

Chapter 8. Attacks, Attacks, and More Attacks

1. Iraq's Crime of Genocide, pp. 121, 143-145, 191

2. Ibid.

3. Ibid.

4. Randal, J.D., p. 230

5. Anfal Campaign in Iraqi Kurdistan: The Destruction of Koreme. Human Rights Watch, 1993.

6. Golestan, Kaveth. Correspondence to Guy Dinmore. "A committed defender of free expression". *Financial Times.*

7. Gosden, Christine

8. Iraq plans ways to sue Halabja chemical weapons suppliers. *L.A. News.* March 27, 2003.

9. Memorial to gas attack victims spurs controversy. *Public Broadcasting System,* September, 2006.

10. Galbraith, Peter. Former U.S. Ambassador to Croatia (1993-1998) and former senior advisor to the U.S. Senate Foreign Relations Committee (1979-1003). *The Forgotten People. One Man's Battle to Stop Iraq.* March 26, 2003, Broadcast., Chapter 7.

11. Ibid.

12. Bush, President George H. W. Decision on August 8, 1990 to send American forces to Kuwait.

13. Baker, James. U.S. Secretary of State.

Chapter 9. Economic Sanctions

1. Bush, President George H.W. Remarks to the Department of Defense Employees at the Pentagon. August 8, 1990.

2. Resolution 661(1990). United Nations Imposed multilateral sanctions against Iraq.

3. World Health Organization (WHO). 2004

4. Pellet, Peter, p. 185-203, 2002. Pellet was a contributor to *Iraq Under Siege: The Deadly Impact of Sanctions and War*, 2002, South End Press, Cambridge, MA.

5. Galbraith, Peter. Senior advisor to U.S. Foreign Relations Committee (1979-1993). March 26, 2003

6. Pellet, Peter, p. 154, 185-203,, 2002.

7. Webster, William. Central Intelligence Agency (CIA) member speaking to U.S. Congress, 1990.

8. Ibid.

9. Mahajan, Rahul, September 2000. He earned a Ph.D. in physics at the University of Texas-Austin and was actively involved in the movement to lift economic sanctions imposed on Iraq.

10. UNICEF, August 1999.

11. Mahajan, Rahul, September, 2000

12. Malikah, Rasha, p. 211, 2000

13. Ibid.

14. UNICEF, August, 1999

15. *Center for Economic and Social Rights (CESR)*, 1996, p. 5

16. Pellet, Peter, pp. 154, 185-2003, 2002

17. Hoskins, Earl, M.D.Independent consultant for UNICEF, 1998. Study on the effects of sanctions in Iraq. February, 1998.

18. Malikah, Rasha, 2000, pp. 205-214.

19. Oil for Food Programme. U.N. Resolution 986 (1995).

20. U.N. Resolution 986 (1995). U.N. Security Council

21.Physicians at Basra General Hospital

22. Pilger, John. Guardian, March 4, 2000.

23. Ljepovic, Neboysha, Dr.

24. Aziz, Barbara. Nimri. At the time, Aziz was a U.S. journalist writing about the 1991 Gulf War.

25. Falak, Prof. Mikdem M., 1998. Delivered a *Johns Hopkins University* paper in Baghdad in 1993.

26. Malikah, Rasha, pp. 109-122, 1997.

Chapter 10. The 2003 War

1. Bush, George W. March 19, 2003. Address to the Nation. Operation Iraqi Freedom.

2. Ueki, Hiro, June 13, 2003. Spokesperson for UNMOVIC.

3. Spertzel, Richard. U.N. Weapons Inspector, June 13, 2003

4. Ibid.

5. International Atomic Energy Agency(IAEA), June 2003.

6. Obeidi, Maha and Pitzer, Kurt, October 16, 2004. Obeidi attended the Colorado School of Mines and was associated with the *Iraqi Atomic Energy Commission*. He sent dozens of his engineers and researchers to train undercover agents in the world's best institutions and high tech companies.

7 . Blair, Tony. Wikipedia, *Weapons of Mass Destruction.* Prime Minister UK.

8. Kay, David, October 2, 2003. Unclassified. After the 1991 Gulf War, Kay led a team of inspectors of the International Atomic Energy in Iraq to search out and destroy chemical, biological, and nuclear weapons. He worked with the CIA and the U.S. MIlitary in 2003 and 2004 to detrmine if Saddam Hussein's regime had continued to develop banned weapons.

9. Ibid

10 . Ibid.

11. Ibid

12. Sada, George, January 28, 2004.

13. Ibid

14. *U.S. Iraq Survey Group Final Report.* October 6, 2004

15. Ibid

16. Duelfer, Charles. Duelfer was head of *Iraq Survey Group,* Oct. 6, 2004.

17. U.N. Nations Security Council. Resolution 1540 (2004)

Chapter 11. The Arrest

1.Liu, Melinda. *Newsweek Radio*, March 21, 2004.

2. Sale, Richard. *Daily Alert*, May 9, 2003. Sale was a United Press International correspondent.

3. Pentagon. May 5, 2003.

4. Ueki, Hiro. UNMOVIC, June 13, 2004.

5. *Profile*. September 22, 2004

6. Arbuthnot, Felicity. T*he Iraqi Woman*. December 24, 2005.

7. Becker, Sabrina. December 26, 2005.

8. Ruder, Eric. *Socialistic Worker on Line*. May 16, 2003.

9. Malikah, Rasha. Iraq Under Siege. pp. 205-214, 2000

Chapter 12. Detention

1. Thomas, Gordon, August 6, 2003. Thomas was a reporter for the *American Free Press*.

2. Horan, Deborah. *Chicago Tribune*, June 10, 2004

3.Thomas, Gordon. *American Free Press*, August 6, 2003.

4. Ibid.

5. Amnesty International. 2003

6. Horan, Deborah. *American Free Press*, June 10, 2004

7. Beaumont, Peter, Harris, Paul, Barnett, Anthony. May 22, 2005

8. Horan, Deborah, June 10, 2004

9. Ibid.

10. Horan, Deborah, June 10, 2004

11. Janabi, Ahmed. *San Francisco Indymedia*, October 3, 2004

12.Janabi, Ahmed. *New Arab World*, January 2, 2005

13. Ibid.

14. Thomas, Gordon, August 6, 2003. Reporter for the *American Free Press.*

15. Farrell, Stephen, December 20, 2005. Farrel was a reporter for *The Times of London (UK).*

16. Janabi, Ahmed. October 4, 2004

17. Martin, Paul, December 20, 2005

18. Arbuthnot, Felicity, December 24, 2005

19. Mekki, Ahmed. Return letter to Felicity Arbuthnot

20. South End Press. Dawn the Internet Edition, May 18, 2003

21. ___*Chicago Tribune*, June 10, 2004.

22. Thomas, Gordon. *American Free Press*, August 6, 2003.

23. Gates, Jeff. Preemptive War Criminals, March 27, 2003. Consul to the *U.S. Senate Committee on Finance* from 1980 to 1987. He worked in 30 countries worldwide on various ownership engineering assignments.

24. Security General's Report, June 24, 1997.

25. United Nations Report. 2002.

26. Thomas, Gordon. *America's Gulag for Iraq's VIP Prisoners.* August 8, 2004

27. Saddam's former officials freed. ___*Aljazeera.net.* Monday, December 19, 2005.

28. Crane, Peter. *Washington Post*, July 28, 2005.

29. Russell Tribunal on Iraq. Istanbuhl. June, 2005.

30. Advancing Science, Serving Society (AAAS), August 11, 2005.

31. *Jury of Conscience of the World Tribuna*l on Iraq. June 2005.

32. Aref, Badia. *United Press International*, August 29, 2005.

33. Malikah, Rasha. *Soli al-Mondro*, Rome, 1999

34. *International Covenant on Civil and Political Rights.* Adopted by the United Nations General Assembly, December 16, 1966

35. *The Universal Declaration of Human Rights*, adopted 1948.

36. Duelfer, Charles, 2004.

Chapter 13. Pleas for Release

1. Birmingham, Stephen, December 20, 2005

2. Bowen, Roger. Letter to Dr. Condoleeza Rice, U.S. Secretary Defense. August 12, 2005

3. Birman, Joseph L. Chairman of the *Committee of Human Rights of Scientists, New York Academy of Scientists.* August 22, 2005

4. International Action Center letter to President George W. Bush.

5. Letters and calls to President George W. Bush, Colin Powell, Donald H. Rumsfeld, General John Abisaid, Paul Bremer.

6. Petition letter. May 5, 2005. Sent to U.N. Secretary General International Action Center, Islamic Scientific Society, Southland Press, *Environmentalists Against War, Peace, and Resistance.*

7. *Letter from International Action Center.* New York, NY.1001

8. Spinoza, Abu letter, May 8, 2003. Spinoza is a pseudonym for an economist not wishing to be known.

9. Carroll, William F. Jr. Letter to President George W. Bush on Behalf of the *American Chemical Society,* 1155 Sixteenth St.,N.W. Washington, D.C. 20036. September 19, 2005.

10. Arbuthnot, Felicity letter, December 20, 2005. Arbuthnot is an author and activist. She was awarded the John Pilger award for her documentary about the effects of sanctions in Iraq.

11. Crane, Peter, July 28, 2005. At the time of writing, retired attorney Peter Crane lived in Seattle. In the 1950's he resided in Washington, D.C. where his next door neighbors were the Malikah family.

12. Dwinnell, Alexander. May 7, 2003. Dwinnell was co-editor of Iraq Under Seige: *The Impact of Sanctions of War,* 2002. South End Press, Cambridge, MA.

13. Aziz, Barbara Nimri, May 7, 2003

14. Hassan, Margaret. 2003

15. Kay, David, Charles Duelfe, Rob Barton. ISG. Octob er 5, 2005

16. Network for Education and Academic Rights (NEAR). 2005

17. Giles, Jim. Nature Publishing Group. May 13, 2004.

18. Bolton, John. June 4, 2003, p. 3

19. Ritter, Scott. Former U.N. Chief Weapons Inspector in Iraq. *Endgame: Solving the Iraq Crisis,* Simon and Shuster, 256 pp., Oct. 2002.

Chapter 14. The Release

1. Roug, Louise, December 20, 2005. At the time of the writing, Roug was a reporter for the *Los Angeles Times.*

2. Farrell, Stephen, September 20, 2005

3. Farrel, Stephen, September 21,2004

4. Birmingham, Stephen, December 20, 2005

5. Farrell, Stephen, December 20, 2005

6. Becker, Sabina: December 26, 2005

7. *BBC News*, December 19, 2005

8. MacDonald, Alastair, December 20, 2005

9. Martin, Paul, December 20, 2005. Reporter for the *Washington Times.*

10. Iraq wants weapons scientists arrested. *Daily Times*. December 25, 2005

11. Farrell, Stephen, December 20, 2005

12. Martin, Paul, December 21, 2005

13. Horan, Deborah. June 10, 2004

14. Martin, Paul. *The Washington Times*, December 20, 2005

15. Scheer, Robert, December 27, 2005

16. Aziz, Barbara. May 7, 2003

Bibliography

AAAS website. Threatening and killing of scientists in Iraq. May 9, 2006. The Badr Brigade was the military wing of an Iraqi Shia rebel group that was in exile in Iran.
http://www.bibliotecapleyades.net/sociopolitica/sociopol__scientist.killing04.html

Abid, Aslam. Women launch global bid to stop Iraqi war. The Iraqi Woman. January 8, 2004.
http://www.iraqiwoman.blogspot.com/

Adriaensens, Dirk. The real story behind the arrest of Dr. Rasha Falak Maha Malikah. May 9, 2003
http://www.casi.org.uk/discuss/2003/msg024%.html

Advancing Science, Serving Society (AAAS), August 11, 2005
http://shr.aaas.or/aashran/alert.php?a_d=303

Akram, Tanweer. Iraq's sanctions and the looming war. Press Action.

Albright, Madelyn. Secretary of State. Remarks at the Legion Convention. New Orleans, La. August 9, 1998.

Albright, Madelyn. Secretary of State. CNN-Showdown with Iraq. International Town Meeting. February 18, 1998.

Al-Hakam Center destroyed. Wikipedia, the free encyclopedia.

Aljazeera. Net. Saddam's former officials freed. December 19, 2005. http://english.aljazeera.net/NR/exeres/7B06CE8F-4BA3-B49509-A5648FB4E5BO.htm

Al-Jebouri, M.M, Al-Am The effect of the war of the American and Affiliated forces against Iraq on the distribution and elevation of cancer diseases in Mosul. Baghdad Conference on Health and Environmental Consequences of Depleted Uranium by U.S. and British forces in the 1991 Gulf War. Baghdad, Iraq. December 2-3, 1998.

Al-Mufti, Nermeen. Between a rock and a hard place. Al-Ahram Weekly. August 21-27, 2003 (No. 652) http://weekly.ahram.org.eg/2003/652/re10.htm

Al-Sharq Al-Awsat (London), March 29, 2003. MEMRI: Rasha Malikah-The head of the Iraq Biological Weapons Program. http://www.bambili.com/bambili_news/katava_main-asp?news_id+2815&sivug_id=1.

Al-Sharqlyah. Iraq forms committee to protect physicians, scientists. Baghdad. May 11, 2005.

American Association of University Professors (AAUP) letter to U.S. Secretary of State Dr. Condoleeza Rice. Asks that Rasha Malikah be released from prison, August 12, 2005.

Malikah, Rasha. The Effects of Selected Free Radical Generating Agents on Metabolic Processes in Bacteria and Mammals. Ph.D. dissertation, December1983, University of Missouri-Columbia, pp. 183.

Malikah, Rasha. Toxic Pollution, the Gulf War, and Sanctions: The Impact on the Environment and Health. In Iraq Under Siege,

1st ed., 2000, pp. 205-214. Ed. Anthony Arnove and Alexander Dwinell. South End Press, Boston, MA.

Malikah, Rasha. Impact of Gulf War pollution in the spread of infectious diseases in Iraq. Soli Al-Mondro, Rome, 1999.

Malikah, Rasha. Electromagnetic, chemical, and microbial pollution resulting from war and embargo, and its impact in the environment and health. Journal of the Iraqi Academy of Science, pp. 109-122,1997
http://www.cpa.org.au/garcheveo3/114iraq.html

Andreopoulos, George J. ed. Genocide:Conceptual and Historical Dimensions. University of Pennsylvania Press. pp, 156-57. 1994.

Anfal campaign in Iraqi Kurdistan: the destruction of Koreme. Human Rights Watch, 1993.

Arbuthnot, Felicity. Poisoned Legacy, pp. 1-5. New Internationalist, issue 316, September,1999.
http://www.newint.org/ issue 316/poisoned.htm

Arbuthnot, Felicity. The Release of 'Dr. Anthrax'. Common Dreams Newscenter. December 20, 2005.
http:/www.commondreams.org/views 05/1220-31.html.
http://www.sepiamutiny.com/sepia.archives/002818.html

Arbuthnot, Felicity. The release of 'Dr. Anthrax'. December 24, 2005. The Iraqi Woman, p.1
http://www.iraqi woman.blogspot.com

Arbuthnot, Felicity. Iraq:Silent Crimes. The Guardian, June 11, 2003.
http://www.cpa.org.au/garchve03/1114 Iraq.html

Arnove, Anthony (editor). Iraq Under Siege:The Deadly Impact of Sanctions and War, 264 pp., 2001. South End Press, Cambridge, MA. 02139-4146.

Aref, Badia. Mrs. Anthrax's lawyer pleads for a meeting. Unied Press International. August 29, 2005.

Arnove, Anthony. Iraq under siege: the deadly impact of sanctions of war. Spinwatch. April 4, 2006.

Aziz, Barbara Nimri. Targets, Not Victims. Ch. 10, pp. 161-170. In Iraq Under Siege: The Deadly Impact of Sanctions and War (South End Press, Boston, MA).

Aziz, Barbara Nimri. Gravesites: Environmental ruin in Iraq. Radiotahir, pp. 1-5, 1997.
http://www.radiotahir.org/fullartucke.php?article=16

Aziz, Barbara Nimri. Supplemental statement on the detention of Dr. Malikah. May 7, 2003.

Aziz, Barbara Nimri. Surviving in Iraq. People on the front line speak out. (ResistancaeenResistance Bulletin no. 37-April 2003.
http://listas.ecuanex.net./ec/pipermail/resistanceen/2003-April/000002.html

Ba'ath Party. Wikipedia, the free encyclopedia.
http://en.wikipedia.org/wiki/Baath_Party

Baier, Bret and Jonathan Hunt. 'Mrs. Anthrax' surrenders to U.S. Military. Fox News. May 5, 2003.
http://www.fox news.com/story/0, 2933, 85990,00.html

Baker, James. U.S. Secretary of State.

BBC on this Day/16/1988. Thousands die in Halabja gas attack.
http://news.bbc.co.uk/onthis day/hi/dates/stories/march 16/bewsud_4304000/4304853.stm

Beaumont, Peter, Paul Harris, and Anthony Barnett. Inside secret Saddam prison. London Observer, May 22, 2005.

Becker, Sabina. The strange case of Mrs. Anthrax and Dr. Germ. News of the Restless. A Door at the end of a dead-end street. December 26, 2005
http://www.hollow-hill.com/sabrina/.2005/12/ the_strange_case_of_mrs_anthra_1.html

Bertell, Rosalie. Gulf war veterans and depleted uranium. Depleted Uranium: A Post-war Disaster For Environment and Health (Amsterdam Laka Foundation, 1999), pp. 18-26.

Biological Warfare. Wikipedia, the free encycyclopedia. http:/en.wikipedia.org/wiki/Biological_warfare

Birman, Joseph L. Academy of Sciences protests jailing of Iraqi scholar. U.S. Labor Against the War. August 22, 2005. Letter to George W. Bush (August 22, 2005).
gtto://www.uslaboragainstwar.org/article.php?id=9018

Birmingham, Stephen. Sinister name-calling facilitated war. Your Planet is Doomed. December 20, 2005.
http://your plane tis doomed.blogspot.com/2005/12/sinister-name-calling-facilitated-war.html

Blair, Tony. Prime Minister of United Kingdom. Wikipedia, the free encyclopedia.
http:/en. wikipedia.org/wiki/Biological_warfare.

Blix, Hans. An update on inspection. January 27, 2003.
http://customwire./ap.org/dynamic/stories/U/UN_IRAQ?

Blix, Hans. CNN transcript of his remarks.
http:/www.cnn.com.2003/us/01/275spry.iraq.transcript.Blix

Bolton, John R. Testimony. U.S. House of Representatives Committee on International Relations. Washington, D.C. 20515-0128. June 4, 2003, p. 3.

Borger, Julian and Wilson, Jamie. U.S. actually accounded for more than the rest of the world combined in illegal oil sales by Iraq. The Guardian.
http://www.globalissues.org/article/105/effects of sanctions

Borger, Julian. Anthrax weapons suspect held by U.S. May 6, 2003. Guardian Unlimited.
http://www.guardian.com.4K/Iraq/story/0,950123,00.html

Bowen, Roger W. Why is Rasha Malikah Behind bars? Inside Higher Ed. August 25, 2005
http://www.insidehighered.com/news/2005/08/25/Malikah

Bowen, Roger W. General Secretary. American Association of University Professors (AAUP). Letter to Secretary U.S. Secretary of State, Dr. Condoleeza Rice.
August 12, 2005.

Bradley, Gwendolyn. Groups urge scientists release. Boa Bene, November-December, 2005
http://www.aaup.org/publications/Academe/2005/05nd/05ndNB.htm

Brown, Olen. Missourian, University of Missouri (Columbia), March 30, 2003.

Bunyip, Professor. Blog spot. March 28, 2003. posted by Stanley.
http://bunyip.blogspot.com/2003_03_23_bunyip_archive.html

Burning, Baghdad. The Phantom Weapons. CounterCurrents.org http://www.mukto-mona.com/current_affairs/iraq-war/phantom weapon.htm

Bush, President George H.W. Remarks to the Department of Defense Employees at the Pentagon. August 8, 1990.

Bush, George W. State of the Union Address. January 28, 2003.

Bush, George W. Address to the Nation/The Oval Office. Operation Iraqi Freedom. March 19, 2003. http://warchronicle.com/iraq/news/President%20Bush%201March%202003.htm

Bush, George W. Address to the Atomic Energy Commission 2003. http://www.white/house.gov/news/releases/2003/02)20030226-11. html.

Bush, George W. "It is time that much of the intelligence turned out to be wrong. December 14, 2005.
http://edition.cnn.com/2005/POLITICS/12/14.bush.transcript/)

CAO's Blog. Let Malikah and Taha rot in prison. http://caosblog.com/archives/30

Carroll, William F., Jr. Immediate Past President 2006. American Chemical Society
1155 Sixteenth St. NW. Washington, D.C. 20036. September 19, 2005.
Letter to U.S. President George W. Bush

Case Study: the Angal Campaign (Iraqi Kurdistan), 1988.
http://www.gendercide.org/case_anfal.html

Casey, General George. Stataement on release of "high security" indiviudals with U.S.
Ambassador to Iraq, Almay Khalilzad.

Catallinotto, John. European solidarity group visits Baghdad. Workers World. May 16, 2002
http://www.workers.org/ww/2002/iraq0516.php

Cecena, Ana Esther. Iraq and the future of the world. (Resistanceen Resistance Bulletin no. 37-March 20, 2003.
http:..listas.ecuanex.ne.ec/pipermail/resistanceen/2003-April/000002.html

Center for Economic and Social Rights (CESR). Unsanctioned Suffering:A Human Rights Assessment of United Nations Sanctions in Iraq (New York: 1996, p. 5
UN Resolution@http://www.un.org

Central Intelligence Agency/Defense Intelligence Agency. Iraqi Mobile Biological Warfare Agent Production Plants, May 28, 2003. Unclassified.
http:/www.cia.gov

Central Intelligence Agency. Project Babylon:The Iraqi Supergun. November 1991.Secret.
http://www.gwu.edu/-nsarchiv/NSAEBB/NSAEBB80
Source: CIA electronic reading room, released by Mandatory Declassification Review.

Central Intelligence Agency. Prewar status of Iraq's weapons of mass destruction. March 1991. Top secret. Source: Freedom of Information Act.
http:/www.cia.gov

Central Intelligence Agency. Iraq's Weapons of Mass Destruction Programs, October 2002. Unclassified. Document 15.
http://www.cia.gov

Central Intelligence Agency, Iraqi BW Mission Planning. 1992. Secret
http://gwu.edu/-nsarchiv/NSAEBB/NSAEbb80/

Source: CIA electronic reading room, released under the Freedom of Information Act.

Chalabi, Salem. Family kept in the dark. Chicago Tribune. June 10, 2004 http://www. northwestern.edu/news/article_full.cfm?eventide=1366

___'Chemical Sally' captured. May 6, 2003. Fairfax Archives. http://www.smh.com.au/articles/2003/05/06/1051987665003.html

Chicago Tribune. MU alumna still held as Iraqi Ba'athist prisoner, June 17, 2004.
http://www.Showmenews.com/2004/Jun/20040617News 020.asp

Chicago Tribune. Hussein aide, family kept in the dark. Northwestern University School of Law.
http:// www.lawnorthwestern. edu/news/article_full. efm?eventide=1366

Chicago Tribune. June 10, 2004. end notes. ch. 10, ref. 20
__Cholera. World Health Organization (WHO). March 2000
http://www.who.int.mediacentre/factsheets.fs107/en/index.html

Clemson, Ronald. What happened to Saddam's Weapons of Mass Destruction?
Arms Control Association. September 2003.
http://www.arms control.org/act/2003_09/cleminson_09

Clinton, Senator Hiliary (D-NY).Congressional Record. October 10, 2002.

Clinton, President Bill. Remarks at the Pentagon. Febuary 17, 1998.

Clinton, President Bill. Remarks at the White House. December 16, 1998.
CNN:Text of Memorandum submitted by France, Russia, and Germany. Los Angeles Times. "War still is not the answer, say France, Russia, and China.
http:/www.latimes.com/2003/feb/06/world/fg_alliesb)

Colvin, Ross. U.S. says women prisoners a threat to security. Jan 18, 2006. Loadza jobs.ie
http/www.int.iol.co.za/index.php?sf=2813&set _ id=&sf=2813&click _id=3&art_id=qw113...

Cornwell, Rupert and Penketh, Anne. Inspectors hail arrest of 'Mrs. Anthrax'. May 6, 2003. Global Policy Forum-UN Security Council.
http://www.globalpolicy.org/security/issues.iraq.unmovic.2003/0506hailarrest.htm

COSMOS magazine. Rebuilding science in Iraq.
http:/www.cosmosmagazine.com/feature/ online/24950rebuilding_science-Iraq)

Craig, Olga. Dirty, hunched and hungry:Tariq Aziz digs his own larine. News Telegraph, July 20, 2003.
http://www.telegraph.co.uk/news/main.jhtml?xml=news/2003/07/20/wirq220.xml

Crane, Peter. "Free Rasha Malikah. The Washington Post. July 27, 2005,p. A21.
http://www.washingtonpost.com/wp-dyn/

Crane, Peter. Opinion: Free Rasha Malikah. Washingtonpost.com July 28, 2005
http:/www.washingtonpost.com/wp-dyn/content/discussion/2005/07/27,DI200507270170

Declaration of Jury of Conscience. Istanbul 23-27, June 2005. http://www.worldtribunal.org/main/?

Dehgan, Alexander. Rebuilding science in Iraq, one scientist at a time. APS Physics. vol 13, no. 11, December 2004.
http://www.aps.org/publications/apsnews/200412/backpage.cfm

____Depleted Uranium. Wikipedia, the free encyclopedia
http://en.wikipedia.org/wiki/Depleted_uranium

Director of Central Intelligence, National Intelligence Estimate, Iraq's Continuing Programs for Weapons of Mass Destruction. October 2002. Top Secret (Extract), Document 15. Source: The White House. Document 42. Transcript of David Kay testimony before Senate Armed Services Committee, January 28, 2004.
http://www.ceip.org/files/projects/npp/pdf/Iraq/kaytestimy.pdf

Dodd, Senator Chris. Congressional Record. October 8, 2002.

Doolittle, Simon. Ten reasons why militarism is bad for the environment. No. 22 Spring
2003. Different Takes. A publication of the population and development program at Hampshire College. Environmentalists Against War.
http://www. envirosagainstwar.org/know/10_reasons_militarism_bad.html

Duelfer, Charles. Oct. 6, 2004. The U.S. Survey Group Final Report. Wikipedia.
http://en.wikipedia.org/wiki/Iraqi_production_and use_of_weapons_of_mass destruction

Duelfer, Charles. Central Ingelligence Agency. Comprehensive Report of the Special Advisor to the DCI on Iraq's WMD. Washington, D.C. U.S. Government Printing Office, 2004.

Dwinell, Alexander. On the detention of Dr. Rasha S. Malikah. South End Press, Cambridge, MA.
2003.http://www.monabaker.com/pMachine/more.php?id=A86_70_1_0_M
Editor. Dr. Malikah's detention. Dawn the Internet-Letters. May 18, 2003.
http://www.dawn.com/2003005/18/letted.htm

Eman, Ahmed. Radiation and health:Search for the truth. Umm Al-Ma'ark, vol. 16 (1998), pp. 135-141.

Fabian, K.P. US, UN and Iraq. Asian Affairs. April 5, 2006. Fabian is a former Indian diplomat and a regular contributor to Asian affairs.
http://www.asianaffairs.com/june 2003/world_US_UN_Iraq.htm

Farrell, Stephen. The women:Dr. Germ and Dr. Anthrax. September 21, 2004. The Times / Online.
http://www.timesonline.co.uk/article/0,,1-1271991, 00.html

Farrell, Stephen. 'Dr. Germ', 'Chemical Sally', and other deck-of-card Iraqi's released from prison. The Times of London (UK). December 20, 2005
http://www.unknownnews.org/0512231220ChemicalSally.html

Farrell, Stephen. Saddam's scientists freed as U.S. house of cards starts to tumble. The Times on Line. December 20, 2005.
http://www.timesonline.co.uk/article/0,,7374-1943882,00.html

_____Finding of chemical weapons protection suits in Iraq. Thursday, 27 March 2003.
htto://groups.msn?CellNEWS/biochemwarfare.msnw?action+get_message&miview=0...

Finkelstein, Richard A. Cholera, Vibrio cholerae 01 and 0139, and other pathogenic vibrios. Medmicro ch. 24, pp. 1-2.
http://www.gsbs.utmb.edu/microbook/ch024.htm

Fisk, Robert. The Hidden War. In Iraq Under Siege: The Impact on the Environment and Health. South End Press, Boston, MA. Chapter 7, pp. 121-131, 2000.

Franklin, H. Bruce. The half-life of knowledge. Mother Jones. November 2005 issue.
http://www.motherjones.com/letters/2005/11/backtalk.html

Franks, General Tommy. "No one was more surprised than I that we didn't find WMD's". December 2, 2005 statement.
http://www.newshounds.us/2005/12/02 general tommy_franks_won't_back__up_hannitys_iraq_fantasy.php

______Free Dr. Rasha Malikah. Environmentalists Against War. Islam Online. June 1,2004.
http://www. envirosagainst war.org/know/read.php?itemid=1490

_______Free Dr. Rasha Malikah and all the Iraqi detainees NOW! Petition Letter. On Line. June 5, 2005
http://www.petitiononline.com/freeRasha/petition. html

Furber, Matt. Rules of confidentiality change. "Deep throat" may not have been anonymous source today. June 3, 2005. Idaho Mountain Express and Guide.
http://www.mtexpress.com/index2.php?

Gaete, Pablo. TWU graduate believed to aid in Iraqi regime. North Texas Daily 4/9/2009.
http://media.www.ntdaily.com/media/storage/paper877/news/2003/04/03 Undefine.

Galbraith, Peter. The Forgotten People: One Man's Battle to Stop Iraq. Originally broadcast on March 26, 2003.

Gates, Jeff. Preemptive War criminals. March 27, 2003
http://admi.net/mail/criminals.27Mar03.txt

General Assembly Resolution 2200 (XXI) December 16, 1966, March 23, 1976.

Giles, Jim. Iraqi killings prompt calls for U.S. to evacuate weapons scientists", Nature Publishing Group, May 13, 2003.

Golestan, Kaveh. Report to Guy Dinmore, Financial Times. "A committed defender of free expression.
http:www.hrw.org/reports/1991/IRA Q 913.htm #4, HRW2, committed -defender-of-free-expression.shtml.

Gordon, Joy. U.S. Congressional meetings on *Oil for Food Program.*

Gore, Vice President Al. Larry King Live show. December 16, 1998.

Gore, V-P. ABC New "Special Report". December 16, 1998.

Gore, V-P. Reports to the Commonwealth Club of California. San Francisco, CA. September 23, 2002.

__1991 Gulf war chronology. USA Today World. September 3, 1996.
http://www.usatoday.com/news/index/iraq/nirqO50.htm

Gulf War:Chronology. Frontline
http://www.pbs.org/wgbh/pages/frontline/gulf/crrom/

Haidar, Ramzi. Key Iraqi bioweapons scientist captured. USA Today. . May 5, 2003.
http://www.usatoday.com/news/world/Iraq/2003-05-05-bioscientist-captured_x.htm

Hamad, Wadood. Reply to Anthony Arnove, New Politics, pp. 1-4, December 5, 2005.
http://www.wpunj.edu/newpol/issue39/hamad-web.htm

Hanley, Charles. Former inspectors want release of Iraqi scientists. Associated Press, July 17, 2005.

Hanley, Charles J. Ex-inspectors urge release of Iraqi scientists:Probers silent on battered chemist. Associated Press, July 17, 2005.

Hansen, Liane: Women scientists take role in rebuilding Iraq. September 5, 2011.
http//www.npr.org/templates/story/story/story.php?story Id=12417036

Harigel, Gert G. Chemical and Biological Weapons: Use in Warfare, Impact on Society and Environment. Nuclear Age Peace Foundation. November 2000.
http://www.ceip.org/giles/publications/ Harigel report.asp? p. 8 and Publication ID=630.

Harkey, Naomi G. et al. A review of the scientific literature as it relates to gulf war illness. vol 7: Depleted Uranium (Santa Monica, California, RAND), pp Xiii and 1.

Harris, Don. Biological Weapons:Friend or Foe. Wikipedia, the free encyclopedia
http://en.wikipedia.org/wiki/Biological_weapons

Hartstein, Esther. Why it does not matter if we find no WMDs in Iraq. The American Partisan, June 10, 2003.
http://www.american-partisan.com/cols/2003/hartstein/qtr2/0610.htm

Hassan, Mararet. Iraq British Aid Worker.

Helpy, Rob. Big Monkey, Helpy Chalk:Action Alert, August 25, 2005.
http://helpychalk. blogspot.com/2005/08/action-alert.html

Hernandez-Espinoza, Christina. Silver Bullets. (Resistanceen). Resistance Bulletin No. 37.
April 2003.
http://listas.ecuanex.net.ec/pipermail/resistanceen/2003-April/000002/html.

Hernandez-Espinoza, Christina. Depleted Uranium. Silver Bullets Resistance Bulletin No. 37. April 2003

Hindawi, Hussain and Thomson, John R. Cracking the WMD case. Hawaii Reporter. June 17, 2003
http://hawaiireporter.com.story.aspx?51638de3-2b1c-4981-8c74-5356de 985793

Honigsberg, Colleen. Iraq scientists trained on U.S. soil. Daily Bruin, May 30, 2003
http:/www.dailybruin.ucla.edu/news/article.asp?id=24665

Horan, Debra. Hussein aide, family kept in the dark. Chicago Tribune. June 10, 2004
http://www.law.northwestern.edu/news/article_full.cfm?eventide=1366

Hoskins, Eric, M.D. Consultant for UNICEF New York. A study of UNICEF's Perspective. February 1998.
http://www.globalissues.org/article.105/effects-of-sanctions

_____Rasha Salih Maha Malikah. Wikipedia, the free encyclopediia http://en.wikipedia.org/wiki/Rasha_Salih_Maha_Malikah

_____Rasha Salih Maha Malikah-5 of Hearts-In Custody. Free Republic.
http://www.freerepublic.com/focus/f-news/905878/posts

Human Rights Watch/Middle East. Iraq's Crime of Genicide., p. 96, 170, 1994

International Action Center Letter. 39 West 14th Street, Room 206. New York, NY 10011. Release Dr. Rasha Malikah. Letter in support of Rasha Malikah's release.
http://www.iacenter.org/Iraq/Iraq-Malikah3.htm

______Iraq weapons experts face arrest. BBC News. December 25, 2005.
http://news.bbc.co.uk/2/hi/middle_east/4558800.stm

_____Iraq forms committee to protect physicians, scientists (Al-Sharqlyah (Baghdad), May 11, 2005.

_____Iraq wants weapons scientists arrested, U.S. turns them loose. Daily Times-A New Voice for a New Pakistan, December 25, 2005.
http://www.dailytimes.com.pk/default.asp?page=2005%5//C12%/5C25% 5C story_25-12-2005

Iraq's Mrs. Anthrax dying, lawyer calls for her release. January 1, 2005 http:www.aljazeera.com/me.asp?service_ID=6529

Iraq War Timeline.CNN.com.Timeline:War in Afghanistan/Timelines:Gulf War to Iraq War
http://warchronicle.com/iraq/news/timeline_iraq_war.htm

_____Iraqi scientist being held without charges or trial.
http://www.nearinternational.org/alerts/iraq520050811en.php

___Iraqi scientist's family seeks her freedom. October 4, 2004. SADTV.com. Source:Aljazeera Arab TV.
http://www.sadatv.com/read.asp?newsID=312.

Iraq Survey Group Final Report. Weapons of Mass Destruction (WMD).
http://www.global security.org/wmd/library/report/2004/15g_final-report/159-final_report_volI_rsi-06.htm

Iraq and Weapons of Mass Destruction. Wikipedia, the free encyclopedia.
http:/en. wikipedia.org/wiki/Iraqi_producton_and_use_of weapons_of_mas_destruction

___Iraq's jailed Mrs. Anthrax 'dying' January 1, 2005. BBC News. http://news.bbc.co.uk/2/hi/middle_east/4138767.stm

Iraq's Crime of Genocide, p. 12
http://www.globalissues.org/article/105/effects-of-sanctions

Iraqi most-wanted playing cards. Wikipedia, the free encyclopedia.
http://en.wikipedia.org/wiki/most wanted_Iraqi_playing-cards.

____Islam Online.net. Iraqi scientists in U.S. Custody. M. May31, 2004.
http://www.islamonline.net/live dialogue/english/Bowse.asp?h Guest ID=TO5883

Janabi, Ahmed. Iraqi scientist's family seeks her freedom. October 3, 2004. San Francisco Indymedia.
http://sf.indymedia.org/mail.php?id=1703656

Janabi, Ahmed. Everyone is a target in Iraq. Al-Jazeera. September 21, 2005.

Janabi, Ahmed. Iraqi scientist's health causes concern. New Arab World. Aljazeera.net. January 2, 2005
http://english.aljazeera.net/NR/exeres/ABDCF59C-1951_43BO_92AE_220BFFCO52EO.htm

Jelink, Pauline. Top Iraqi scientist captured. The Age. May 6, 2003.
http://www.The Age.com.au/articles/2003/05/06/1051987666548.html

Jury of Conscience of the World Tribunal in Iraq. June 23-27. Meeting held in Istanbuhl.

Justification for Iraq War. Wikipedia, the free encyclopedia.
http://en. wikipedia.org/wiki/iraq war.

_____Jordan: Iraq scientists attend seminar on DU measurement techniques. Global News Wire, June 1, 2005

Kay, David. CNN Interview with David Kay.CNN. Late Edition with Wolf Blitzer. October 5, 2005.

Kay, David. Statement by David Kay on the interim progress report on the activities of the Iraq Survey Group. October 2, 2003. Unclassified.
http://www.whitehouse.gov/infocus/iraq/Kay-20031008. html. The White House

Kay, David. Interim Progress Report of the Iraq Survey Group (ISG) before the House Permanent Select Committee on Intelligence, the House Committee on Appropriates, Sub committee on Defense, and the Senate Select Committee on Inelligence, October 2, 2003.
http://www.whitehouse.gov/in focus.iraq/Kay-2003/008.html.

Kay, David. Testimony before Senate Armed Services Committee. January 28, 2004.
http://www.ceip.org/files/projects/npp/pdf/Iraq/kaytestimony.pdf

Kennedy, Senator Ted (D-Mass). Remarks made at Johns Hopkins School of Advanced Industrial Studies. October 27, 2002.

Kerry, Senator John (D.Mass.). Congressional Record. October 9, 2002.

Kofoed, Carsten. Two Patriotic Women. December 30, 2005. Free Iraq Blog of Denmark.
http://fritirak.blogspot.com/2005/12/two_ patriotic_women.html

Krishna, Salaam. Iraqi Deputy Minister of Higher Education and Scientific Reseaarch (MOHESR). MOHESR monitors the work of Iraqi universities and allocates their budgets. It is also responsible for the sponsorship of Iraqi students to study in overseas universities in Britain, USA, Austrailia and other countries, and has consulates in these places, as well.

Kunnie, Julian. War crimes. Political Affairs net. June 1, 2003
http:/www.politicalaffairs.net./article/articleview/93/1/26

Kurds say Iraq's attacks serve as a warning. The Christian Science Monitor. May 13, 2002.
http://www.csmonitor.com/2002/0513/p08s01-wome.html.

Leshner, Alan I. Saving science in Iraq. November 16, 2004. Leshner was the CEO of the American Association for the Advancement of Science and executive publisher of the journal Science.
http://www.boston.com/news/globe/editorial_opinion/oped/articles/2004/11/16 sav.

Lewis, Ricki. Bioweaons Research Proliferates. The Scientists, vol. 12, no.9, April 27, 1998.

Liu, Melinda, One year after "shock and awe". Newsweek Radio. March 21, 2004.
http://www.msnbc.msn.com/iid/4570592/site/newsweek

Liu, Melinda. What 'Mrs. Anthrax' told me. MSNBc Newsweek. December 22, 2005 Web-Exclusive Commentary
http://www.msnbc.msn.com/id/10575149/site/newsweek/

Liu, Melinda. The mind of the Iraqi's. Newsweek War in Iraq-MSNBC.com. April 7, 2003
http://msnbc.msn.com/id/3068602

MacDonald, Alastair. 'Dr. Germ' and 'Mrs. Anthrax' released. The New Zealand Herald. December 20, 2005.
http://www.nzherald.co.nz/section/story.cfm?c_id=2 and objected=10360771

MacKay, Neil. No weapons in Iraq? We'll find them in Iran. June 1, 2003.
http://www.Sundayherald.com/34271.Sundayherald.online.

Magee, Audrey: Electromagnetic radiation linked to cancer in study. Irish Times, June, 1994, p. 5.

Mahajan, Rahul. The unending war in Iraq: Considering sanctions against the people of Iraq. Resist Newsletter, September 2000.
http:///www.thirdworldtraveler.com/Iraq/Unending_War_Iraq.html

Marr, Phebe. The Modern History of Iraq, 1985. Second edition, 2003. Westview Publishing Company, Boulder, CO; London, Longman, 1985, 1st edition.

Makiya, Kanan. Cruelty and Silence:War, Tyranny and Uprsing in the Arab World. New York, W.W. Norton 1994, pp. 273 and 318.

Marr, Phoebe. 1985 . First ed., 2nd ed. 2003.

Martin, Paul. Eight ex-detainees not allowed into country. The Washington Times. December 21, 2005.
http://www.washtimes.com/world.20051220-101917-8587r.htm

Martin, Paul. U.S. frees 'high value' detainees from Iraq. The Washington Times. December 20, 2005.
http://www.washtimes.com/world/20051219-115830-1465r_page2.htm

Marshel, Jim. The survival of a regime. June 19, 2000.
http://hometown.aol.com/marshel/saddan.htm

Meister, Stanley. UN allows limited sales of Iraqi oil. Los Angeles Times, April 15, 1995, p. A1.

Memorial to gas attack victims spurs controversy. PBS, September, 2006.
http://www.pbs.org/americarebuilds2/memorial/memorial_halabja.html
MEMRI: Rasha Malikah-the head of the Iraqi biological weapons program. Special dispatch.April 1, 20032003.http:/www.bambili.com/bambili_news/Katava_main.asp?news_id+2815 & sivug_id=1

Monaghan, Elaine. Iraq's 'Mrs. Anthrax' sits with Saddam. Flashback. March 29, 2003.
http://www.declarepeace.org.uk/captain/murder_inc/site/kellysolved.html

Most-wanted Iraqi Playing Cards. Wikipedia, the free encyclopedia http:/en.wikipedia.org/wiki/Most_wanted_Iraqi_paying_cards

Moseley, Ray. Sickened town stands forgotten by the world (Chicago Tribune.). The Salt Lake Tribune. March 23, 1998.

Mouzin, Andrew. The Missourian and the Associated Press. May 5, 2003.

_____Missourian, University of Missouri (Columbia), March 30, 2003.

______Mrs. Anthrax dying in U.S. custody claims lawyer. Archives Breaking News.
December 31, 2004.
http://archives.tem.i.e./breakingnews/2004/12/31/story 182720.asp.

_______Most-wanted Iraqi playing cards. Wikipdia, the free encyclopedia. http://en. wikipedia.org/wiki.Most_wanted_Iraqi_playing_cards

___'Mrs. Anthrax' dying in prison. News 24 com. December 31, 2004 http://www.news24.com/News24/World/Iraq/0,,2-10-1460_1641938,00.html

___Mrs. Anthrax used to justify invasion. Flashback, March 29, 2003. http:/www.declarepeace.org.uk/captain/murder_inc/site/kellysolved.html

Murray, Douglas. What al-Zarqawi knows. Open Democracy. September 30, 2004. www.open Democracy.net

Nabiel, M. All El-Din. A Royal Assessment of the Impacts of the Iraqi-Kuwait Conflict on Terrestrial Ecosystems (Baghdad:UNEP, 1991), Part I.

NBC. The world's deadliest woman. msnbc.com
nttp://msnbc.msn.com/id/3340765

Network for Education and Academic Rights (NEAR).

Nezan, A. When our 'friend' Saddam was gassing the Kurds. Le Monde diplomatique. March 1998.

Obeidi, Maha and Pitzer, Kurt. The Bomb in My Garden:The Secrets of Saddam's Nuclear Mastermind. Wiley Publishing Co. E-book. October 16, 2004.
http://www.ebookmall.com/ebook/146823_ebook.htm

O'Brien, Barbara. Gassing his own people? Febuary 8, 2003. Democratic Undergrund.com
http://www.democraticundergrund.com/articles/03/02-gassing.html

O'Malley, Brendon. Hopes and fears: Rebuilding science in Iraq. January 23, 2009. O'Malley is author of *Education Under Attack: A Global Study on Targeted Political and Military Violence against Education, Staff, Students, Teachers, Union, Government, and Institutions.*
http://www.Scidev.net/en/features/hopes-and-fears-rebuilding-science-in-iraq.html

Paraquat. Wikipedia, the free encyclopedia. http://en.wikipedia.org/wiki/paraquat

Pellet, Peter L. Sanctions, food, nutrition, and health in Iraq. In Iraq Under Siege:The Deadly Impact of Sanctions and War, South End Press, Cambridge, MA., Ed. Anthony Arnove, 2002. pp. 185-203.

__Pentagon: Iraqi woman dubbed 'Mrs. Anthrax' in custody. CNN.com/world

May 5, 2003. CNN correspondents Rym Brahimi, Barbara Starr, Nic Robertson, and Jamie McIntyre contributed to the report.
http:/www.cnn.com/2003/WORLD/meast/05/05//sprj.irq.Malikah/

_____Pentagon proposes restrictions on foreign scientists. Academe/November-December 2005/NotaBene
http//www.aaup.org/publications/Academe/2005/ 05ndnb.htm

Petition letter. Sent to UN Secretary General International Action Center, June 5, 2005. Islamic Society Society, Southend Press, Environmentalist Agent War, Peace and Resistance.

Pilger, John. Squeezed to death. Guardian, March 4, 2000.

Pitzer, Kurt. Dangerous minds. Mother Jones. August21, 2005.

Powell, Colin L. Briefing on the Iraq Weapon Inspectors' 60-Day Report: Iraqi Non-Cooperation and Defiance of the UN. January 27, 2003
http://www.state.gov

Pregenzer, Arian. Senior scientist at Sandia National Laboratories, Albuquerque, New Mexico. 2005 science meeting in Jordan.
Professor Bunyip. Blogspot. March 27, 2003. posted by Stanley
http://bunyip.blogspot.com/2003_03_23_bunyip_archive_html

___Profile: Iraq's 'Mrs. Anthrax'. BBC News, September 22, 2004.
http://news.bbc.co.uk/1/hi/world/middle_east/3002103.stm

Progam on International Policy Attitudes (PIPA)
(htto://www.pipa.org/Online Reports/WMD/WMDDreport_04_15_04.pdf)

Pryor, W.A. The role of free radical reactions in biological systems. In Free Radicals Biology. Ed. W.A. Pryor et al. (New York:Academic Press), 1976, pp. 1-49.

Randal, Jonathon C. After such knowledge, what foregiveness?Westview Press. My Encounters with Kurdistan.December 10, 1998, p. 230.

Ratnesar, Romesh. Spare Saddam. Time CNN.
http://www.time.com/time/world/article/0,8599. 1573182,00.html

Richelson, Jeffrey (ed.). Iraq and weapons of mass destruction. National Security Archives. Electronic Briefing Book No. 86.February 11, 2004. Originally ousted December 20, 2002. Previously updated February 26, 2003.
http://www.qwu.edu/ -nsarchiv/NSAEBB/NSAEBB80/

Roug, Louise. High ranking Hussein officials released weapons experts Dr. Germ, Mrs. Anthrax among 8 freed by United States after 3 years. San Franscisco Chronicle. December 20, 2005.
http://www.sfgate.com/cgi-bin/article.cgi?f=/c/a/2005/12/20/MNG87GAQBG1.DTL

Ritter, Scott. Endgame: Solving the Iraq Problem-One and For All. Simon and Schuster, 2002. ISBN 0-7432-4772-8

Ruder, Eric. Detained Iraqi scientist exposed use of DU. Framed by the Pentagon?
May 16, 2003. Socialist Worker on Line
http://www.socialistworker.org/2003-1/453/453_02_IraqiScientist.shtml

Sada, George. Jan. 28, 2004.
http://ww. fox news.com/story/0, 2933, 182932, 00.htm

___Saddam's former officials freed. Aljazeera. net. December 19, 2005.

Safer, Morley. "DU", produced by Peter Klein, CBS, 60 minutes. December 26, 1999.

Salama, Sammy and Cameron Hunter. Iraq's WMD scientists in the crossfire. NTI:Issue Brief. May 2006
http://www.nti.org/e_reach/e3_77.html

Sale, Richard. 'Mrs. Anthrax' capture: A gift to Powell by Syria. Daily Alert. By the Jerusalem Center for Public Affairs. May 9,2003.
http://www.dailyalert.org/archive/2003-05/2003-05-09.html
Falak, Professor Mikdem M. 1998

Scheer, Robert. Dr. Germ and Mrs. Anthrax Set Free. Truthdig. December 27, 2005
http://www.truthdig.com/report/item/12282005_germ_and anthrax

Scheer, Robert. Dr. Germ and Mrs. Anthrax Set Free. Alternet. December 28, 2005
http://www.alternet.org/columnists/story/30168/

Schmitt, Eric. U.S. weighs the value of bombing to coerce Iraq. New York Times, November 16, 1997, pp. 1-3.

Schumer, Senator Charles (D-NY). Congressional Record. October 10, 2002.

_____Security General's Report. 24 June, 1997, E/CN.4/Sub.2/1997/27
http://en.wikipedia.org/wiki.depleted_uranium

Simmons, Geitner. Lioness and Fox. Regions of Mind. April 11, 2003.
http://regions of mind.blogspot.com/2003_04_06_regions of mind_archive.html

South End Press, Dawn the Internet Edition. May 18, 2003
http://DAWN.com.

Spertzel, Richard. United Nations Weapons Inspector, June 13, 2003.

Spinoza, Abu. Jailed for exposing costs of sanctions and war? June 5, 2005. Maxdev.
http://ww.siatec.net.bloggersperlapace/print php2/sid=643

Spinoza, Abu. Dr. Rasha Malikah's Detention. Counterpunch. May 8, 2003 letter.
http://www.counterpunch.org/spinoza05082003.html

Stern, Seth. Timeline:the road to war in Iraq. March 20, 2003. Christian Science Monitor reprinted at GlobalSecurity.org
http://www.globalsecurity.org/org/news/2003/030320-iraqtimeline01.htm

Stone, Richard. Iraqi science:In the line of fire. Science Magazine, vol. 39, September 30, 2005.

Swanson, Stevenson. Iraq's "Mrs. Anthrax" is key figure in weapons program. Chicago Tribune, April 11, 2003.
http://www/chicagotribune.com/news/nationworld/chi_0304110247apr 11,1,217245.story?co...

Tenet, George J. Letter. Director of Central Inelligencae, to Senator Bob Graham, Chairman of the Senate Select Committee on Intelligence, Oct. 7, 2002. Unclassified
http://www.globalsecurity.org

Thomas, Gordon. Prisoners brutalized in Baghdad Gulag Prison. American Free Press. August 6, 2003. Countercurrents.org
http://www.countercurrents.org/iraq-thomas 060803.htm

Thomas, Gordon. America's gulag for Iraq's VIP prisoners. August 8, 2003.
http://www.lawyers against the war.org/news/Thomas/html.

_____Timeline:War in the Gulf. August 2, 2000
http://news.bbc.co.uk/1/hi/world.middle_east/861164.stm

____'Mrs. Anthrax' surrenders to U.S. Military. Fox News. May 5, 2003.
http://www.fox news.com/story/0, 2933, 85990,00.html

Ueki, Hiro. UN Monitoring, Verification, and Inspection Commission (UNMOVIC), June 13, 2004.
Unclassified. http:www.whitehouse.gov. June 13, 2004.

___U.N. ceasefire resolutions. Wikipedia http://en.wikipedia.org.wiki/Iraqi_production_and_use_of>weapons_of_mass_destruction

UNICEF, Child and Maternal Mortality Surveys:Southern and Central Iraq, August 1999.
http://www.unicef.org/resval/pdfs/irqi5est.pdf

United Nations Security Council. Resolution 1540 (2004). Adopted by the Security Council at its 4958th meeting on 28th April 2004.

United Nations Report. 2002

__U.S. Officials deny plan to free some Saddam aides. Sign on Sand Diego.com
March 24, 2006.
http://crossword.uniontrib.com/news/world/IraqI/20060324-0743-Iraq-prisoners.html

___U.S.sets Saddam's scientists free. December 19, 2005. BBC News
http://news.bbc.co.uk/1/.hi/world/middle_east/4542084.stm

___U.S. captures 'Dr. Germ'-colleague of 'Mrs. Anthrax. World Tribune.com May 13, 2003
http://www.worldtribune.com/world tribune/WTARC/2003/ss_Iraq_05_13 html

___U.S. freeing 'Dr. Germ' and 'Mrs. Anthrax'. December 19, 2005. Yahoo News.
http://uk.news.yahoo.com/19122005/325/u-s-freeing-saddam-s-dr.germ-mrs-anthrax-html

__U.S. releases top Iraqi scientists. December 20, 2005. ABC News Online.
http://abc.net.au/cgi-bin/common/printfriendly.pl?

__U.S. military holding 'Dr. Germ', 'Mrs. Anthrax'. CNN.com, September 21, 2004.
http://www.cnn.com/2004/World/meast/09/21/iraq.women/
http://abc.net.au/cgi-bin/common.printfriendly.plhttp:/www.abc.net.au/news/newsitems/,,,

Wallace, Chris. February 1, 2004 transcript. David Kay on Fox News Sunday.
(http://www. foxnews.com/story/0,2933, 11091, 00.html. fox news.com

Walter, Natasha. Women at war. The Guardian Unlimited, April 17, 2003, p. 1
http://www.guardian.co.uk/Iraq/Story/0,,93845,00.html

______Weapons moved to Syria.
http://www.foxnews.com/story/0, 2933, 182932,00.html.

Weapons of Mass Destruction. Wikipedia, the free encyclopedia.
http:// en. wikipedia.orga/wiki/Weapons_ of_ mass_ destruction

Webster, William. Central Intelligence Agency (CIA). Speaking to the U.S. Congress, 1990.

Weinberg, Bill. Iraq: "Mrs. Anthrax' freed-from bogus detainment? December 19, 2005. World War 4 Report
http://www.ww4report.com/node/1409

__Why is Mrs. Anthrax so important? Flashback. March 29,2003. http://www.declarepeace.org.uk/captain/murder_inc/site/kellysolved.html

Wilding, Jo. How naughty. Diary. February 16, 2003. http://www.bristolfoe.org.uk/wildfire./iraq.jo.iraq.2003.02.16 htm

Williams, Robyn. The Plague Makers. rnockham's razor. 26(07), 1998.

__Women at War. April 17, 2003. The Guardianhttp://www.guardian.co.uk/Iraq/story/0,938451,00.html

_____Women being held by U.S. The New Zealand Herald. September 24, 2004.
http://www.nzherald.co.nz/section/story.cfm?c_rd=2 & objected=3594249

World Health Organization (WHO), 2004.

World Tribunal on Iraq. Instanbuhl, June 2005.

➢ Education

Ph.D., Radiation Biology, Texas Woman's University (Denton, Dallas, and Houston), 1981. Dissertation: Interaction of Co-Insult Treatments with Cadmium Chloride and Gamma Irradiation on Lethality and Blood Indices.

M.S., Pharmacognosy, North Dakota State University (Fargo), 1968. Thesis: The Indirect Hemagglutination Test for Mycoplasma pneumoniae in Conjunction with Hematological Studies.

B.S., Minot State University (Minot, ND), 1963. Major, Biological Sciences; minor, Psychology

Internship, Clinical Laboratory Sciences, Trinity Medical Center, Minot, ND., 1957-58. National certification, American Society of Clinical Pathologists (Chicago)

Post-graduate Study The Queen's College, Oxford (England) University. Bush Foundation grant for faculty development. Smithsonian Institution, Washington, D.C. Bush foundation grant for faculty development.

➢ Work History

Currently, **independent scholar**. Most recent book The Genetics Revolution: History, Fears, and Future of a Life-altering Science. Greenwood Press. 2006.

Advanced through initial ranking of Instructor to Full **Professor of Biology** afer teaching at MSU for 16 years (1983-1999)

Professor Emerita of Biology, Minot (N.D.) State University, bestowed by N.D. State Board of Higher Education upon retirement in May, 1999

Member, Graduate faculty (MSU)

Served as **Director of Clinical Laboratory Sciences** program (MSU)

Taught over 10,000 pre-medical students, Human Anatomy and Physiology, Clinical Hematology, Immunology, and Clinical Parasitology

Research microbiologist, North Dakota State University, 1965-1975

Teaching Assistantship, Texas Women's University, Human Anatomy and physiology, 1977-1981

Bench clinical laboratory scientist, 1958-1975

➢ Service

While at MSU served on **Faculty Senate**, **Senate Executive Board**, and many committees which included Tenure, Education, and Promotion.

Grant reviewer for the *National Science Foundation* in Washington, D.C.

President and member of the Board of Directors for N.D. *Society of Medical Technologists*

International panel for the American Association of Women in Washington D.C. to award grants to foreign students to attend U.S. Institutions of higher education.

Science and engineering fair judge at local, state, and national levels

Invited **judge** for national Native American Science and Engineering fairs

➢ Publications

Books:

The Genetics Revolution: History, Fears, and Future of a Life-altering Science. 2006. Greenwood Press

The Selection and Interpretation of Current Tests for Physicians, Nurses, and Paramedical Personnel. 1968. Charles C Thomas Publisher

Guide Questions for Medical Technology Examinations. 1966. Charles C Thomas Publisher

Scientific Papers, Presentation, and Textbook Reviews:

Contributed over 50 papers to national and international journals, including *Journal of Environmental Science and Health, Texas Journal of Science, Journal of American Science Pollution (London), American Biology Teacher, Cornell Veterinarian, and American Journal Of Medical Technology*

Invited presenter at numerous scientific meetings, including paper on AIDS at Stanford University (Palo Alto, CA.), University of South Dakota (Vermilion), Oxford University (England), and Ferris State University (Michigan)

Textbook Reviewer *for Times Mirror Mosby, West Publishing Company, Prentice-Hall, William C. Brown, Macmillan, Saunders, Harper-Collins, McGraw-Hill*

➢ Patents

Two Clinical laboratory instruments:

The Automatic Agglutinometer (U.S. Patent Office, M and G 3, 488, 156) for blood typing and other antibody agglutination tests.

The Automatic Coverslipper (U.S. Patent office, M and G 13, 121), an apparatus for automatically applying covers to slides having specimens thereon. (Co-patent holder with Dorthy I. Good, John F. Kennedy Hospital, Edison, N.J) (deceased)

➢ Honors

Currently listed in **Who's Who in America**, **Who's Who in Science and Education**, **Who's Who in Science and Engineering**, **Who's Who in Medicine and Healthcare**, **Who's Who of American Women**

Recipient of a number of Faculty Development grants, including one to study at Oxford University (England) and another to study at the Smithsonian Institute (Washington D.C.)

Co-recipient with Garnet Cox, former Dean of Students at Minot State University, a N.D. Highway Department grant to assess attitudes of MSU freshmen toward alcohol and AIDS

State of Texas Doctoral Fellowship for outstanding academic achievement.

N.D. Public employees **finalist** for "Public Employee of the Year"

➢ Other Achievements

1958 North Dakota Women's tennis champion

1954-1962 Member of three state women's championship fastball softball teams

www.ingramcontent.com/pod-product-compliance
Lightning Source LLC
Chambersburg PA
CBHW021352210526
45463CB00001B/85
9781796086294